Studies in Systems, Decision and Control

Volume 289

Series Editor

Janusz Kacprzyk, Systems Research Institute, Polish Academy of Sciences,
Warsaw, Poland

The series "Studies in Systems, Decision and Control" (SSDC) covers both new developments and advances, as well as the state of the art, in the various areas of broadly perceived systems, decision making and control–quickly, up to date and with a high quality. The intent is to cover the theory, applications, and perspectives on the state of the art and future developments relevant to systems, decision making, control, complex processes and related areas, as embedded in the fields of engineering, computer science, physics, economics, social and life sciences, as well as the paradigms and methodologies behind them. The series contains monographs, textbooks, lecture notes and edited volumes in systems, decision making and control spanning the areas of Cyber-Physical Systems, Autonomous Systems, Sensor Networks, Control Systems, Energy Systems, Automotive Systems, Biological Systems, Vehicular Networking and Connected Vehicles, Aerospace Systems, Automation, Manufacturing, Smart Grids, Nonlinear Systems, Power Systems, Robotics, Social Systems, Economic Systems and other. Of particular value to both the contributors and the readership are the short publication timeframe and the world-wide distribution and exposure which enable both a wide and rapid dissemination of research output.

** Indexing: The books of this series are submitted to ISI, SCOPUS, DBLP, Ulrichs, MathSciNet, Current Mathematical Publications, Mathematical Reviews, Zentralblatt Math: MetaPress and Springerlink.

More information about this series at http://www.springer.com/series/13304

B. V. Senthil Kumar · Hemen Dutta

Multiplicative Inverse Functional Equations

Theory and Applications

Springer

B. V. Senthil Kumar
Department of Information Technology
Nizwa College of Technology
Nizwa, Oman

Hemen Dutta
Department of Mathematics
Gauhati University
Guwahati, Assam, India

ISSN 2198-4182 ISSN 2198-4190 (electronic)
Studies in Systems, Decision and Control
ISBN 978-3-030-45357-2 ISBN 978-3-030-45355-8 (eBook)
https://doi.org/10.1007/978-3-030-45355-8

This Springer imprint is published by the registered company Springer Nature Switzerland AG
The registered company address is: Gewerbestrasse 11, 6330 Cham, Switzerland

Preface

The study of solutions and stability results of functional equations is a hot topic in the research field of analysis. The occurrence of many functional equations can be found in various areas such as information theory, probability, dynamic programming, image processing, physics, chemistry, computer graphics, economics, geometry, social science, population ethics, theory of relativity, combinatorics, communication and wireless networks, optics, and cryptography. The stability results of functional equations are employed in non-linear analysis, especially in fixed point theory. The stability results are used to study asymptotic properties of additive mappings. These interesting concepts motivated us to bring the stability results of some new multiplicative inverse functional equations in the form of a book. The book introduces many multiplicative inverse functional equations and their stability results in various spaces. Counter-examples have been given when the stability results fail for singular cases. The book should be a valuable resource for researchers, graduate students, and teachers interested in functional equations, and should also be useful for seminars in analysis covering topics of functional equations. It consists of seven chapters, and they are organized as follows.

Chapter 1 aims to impart the significant role of functional equations in various fields. The study of functional equations is a growing and an important area in mathematics. It covers many other areas of mathematics and recently their role in science and engineering has become very attractive to the researchers. In this connection, we portray few applications of functional equations in geometry, finance, information theory, wireless sensor networks and electric circuits with parallel resistances.

Chapter 2 deals with the investigation of validity of various fundamental stabilities of multiplicative inverse type tredecic and quottuordecic functional equations relevant to Ulam stability theory in non-Archimedean fields via fixed point method. Two suitable counter-examples are also included to prove that the stability results are not valid for singular cases.

Chapter 3 is devoted to demonstrate the validation of various stabilities of multiplicative inverse quindecic and multiplicative inverse sexdecic functional equations via fixed point technique in the framework of Felbin's type fuzzy normed spaces. Proper illustrations are presented to disprove the stability results for singular cases.

Chapter 4 contains the classical investigation of various fundamental stability results of multiplicative inverse septendecic and octadecic functional equations in quasi-β-normed spaces using fixed point technique and also includes two proper examples to disprove stability results for control cases.

Chapter 5 is devoted to study various classical stability results of multiplicative inverse novemdecic and vigintic functional equations in intuitionistic fuzzy normed spaces and also counter-examples to disprove the validity of stability results for singular cases.

Chapter 6 aims to present the generalized Hyers-Ulam stability of multiplicative inverse unvigintic and duovigintic functional equations in paranormed spaces using direct and fixed point methods. Counter-examples to invalidate the stability results for critical cases are also discussed.

Chapter 7 aims to achieve an inexact solution near to the exact solution of a multiplicative inverse trevigintic and quottuorvigintic functional equations in the sense of Ulam stability hypothesis in matrix normed spaces. Proper examples are also illustrated to prove the instabilities for control cases.

The authors would like to thank all the mathematicians who have dealt with several functional equations in the literature available so far. The authors also thank their family members, friends and well-wishers who encouraged them to bring out this book. The authors sincerely welcome productive suggestions and comments to improve the quality of the book for the next edition.

Nizwa, Oman   B. V. Senthil Kumar
Guwahati, India   Hemen Dutta
February 2020

Contents

About the Authors

Dr. B. V. Senthil Kumar is serving in the Department of Information Technology, Nizwa College of Technology, Nizwa, Oman. His areas of interest are solution and stability of Functional, Differential and Difference equations, Operations Research, Statistics, and Discrete Mathematical Structures. He obtained his Ph.D. Degree in 2015. He has 20+ years of Teaching and 10 years of Research experience. He has published more than 50 research papers in reputed peer-reviewed indexed National and International Journals. He has co-authored four books on different titles and contributed three chapters in books. He has delivered invited talks in various institutions and also organized many academic and non-academic events. He is a member of many mathematical societies and a member of editorial committee for several journals. He is a reviewer of many international journals.

Dr. Hemen Dutta is a regular faculty member in the Department of Mathematics at Gauhati University, India. His research areas include mathematical analysis, mathematical modeling, etc. He has to his credit over 100 items as research papers and chapters in books. He has published 14 books as textbooks, reference books, monographs, edited books and conference proceedings. He has delivered several talks at national and international levels and organized several academic events in different capacities. He has also published several articles in the newspaper, popular books, magazines and science portals.

Chapter 1
Introduction to Functional Equations and Ulam Stability Theory

Abstract In this chapter, we impart the significant role of functional equations in various fields. The study of functional equations is a growing and an important area in mathematics. It covers many other areas of mathematics and recently the role of functional equations has become very attractive to the researchers in science and engineering. We portray few applications of functional equations in geometry, finance, information theory, wireless sensor networks and electric circuits with parallel resistances.

1.1 Significance of Functional Equations

In mathematics, a functional equation is any equation that specifies a function in implicit form. Often, the equation relates the value of a function (or functions) at some point with its values at other points. For instance, properties of functions can be determined by considering the types of functional equations they satisfy. The term functional equation usually refers to equations that cannot be simply reduced to algebraic equations.

The investigations of functional equations in other subjects like Differential Geometry, Iterations and Analytic Functional, Differential Equations, Number Theory, Abstract Algebra indicate the growing importance of functional equations. In this way this theory acquired its own personality. The reason for this interest in functional equations by the mathematicians of all the world is due to the fact that in many branches of the mathematics, analytical methods have already been exhausted to some extent. A use of elementary methods often allows one to obtain much deeper and more general results than it was possible with a use of classical methods of Mathematical Analysis.

Functional equations arise in many fields of Mathematics, such as Mechanics, Geometry, Statistics, Measure Theory, Algebraic Geometry, Group Theory. Functional equations find many applications in the study of stochastic process, classical mechanics, astronomy, economics, dynamic programming, game theory, computer graphics, neural networks, digital image processing, wireless sensor networks,

B. V. Senthil Kumar and H. Dutta, *Multiplicative Inverse Functional Equations*, Studies in Systems, Decision and Control 289, https://doi.org/10.1007/978-3-030-45355-8_1

statistics, information theory, coding theory, fuzzy set theory, decision theory, multivalued logic, artificial intelligence, cluster analysis, multivalued logic, population ethics, financial management, geometry, electric circuits, probability theory, cognitive science, iteration, dynamical systems, psychometry, nondifferentiable functions, binomial expansion, scalar products, economics, the Cobb-Douglas production function and quasilinearity, interest formulae, population ethics, Gaussian functions, Chebyshev polynomials, determinants and sums of powers and many other fields. Functional equations are being used with vigor in ever-increasing numbers to investigate problems in the above-mentioned areas and other fields.

It is impossible to provide an informative survey of all the works on the various applications of functional equations, and so we provide some remarkable applications of functional equations.

1. The following two-variable functional equation

$$C_1(x, y)P(x, y) = C_2(x, y)P(x, 0) + C_3(x, y)P(0, y) + C_4(x, y)P(0, 0)$$

   where $C_i(x, y)$; $i = 1, 2, 3, 4$ are given polynomials in two complex variables x, y, arises from different communication and networks systems [52].
2. The Cauchy additive functional equation $f(x + y) = f(x) + f(y)$ is used in genetics to find the combinatorial function $g_r(n) =$ the number of possible ways of picking r objects at a time from n objects allowing repetitions, since this function describes the number of possibilities from a gene pool. For further details, one can refer [68].

3. Using logarithmic Cauchy functional equation $g(xy) = g(x) + g(y)$, it is easy to prove that $\int_1^x \frac{1}{t} dt = \ln x$ [130].
   Let us define $\phi : \mathbb{R}_+ \to \mathbb{R}$ by

$$\phi(x) = \int_1^x \frac{1}{t} dt, \qquad x > 0.$$

Hence, in the case $x, y \in (1, \infty)$, we have

$$
\begin{aligned}
\phi(x) + \phi(y) &= \int_1^x \frac{1}{t} dt + \int_1^y \frac{1}{t} dt \\
&= \int_1^x \frac{1}{t} dt + \int_x^{xy} \frac{1}{z} dz, \qquad \text{where} \quad z = tx \\
&= \int_1^{xy} \frac{1}{w} dw \quad \text{(additive property of the integral)} \\
&= \phi(xy).
\end{aligned}
\tag{1.1}
$$

We present some examples to illustrate how functional equations are applied to solve some interesting problems in Geometry, Finance, Information theory, Wireless sensor networks and electric circuits with parallel resistances.

1.2 Application of Functional Equation in Geometry

Area of Rectangle: In 1791, Legendre applied functional equations to obtain the area of a rectangle. Consider the rectangle whose base is b and height is a. We are interested in finding the area of the rectangle. Let us assume that the area of the rectangle is $f(a, b)$.

Now, divide the rectangle horizontally so that the rectangle is divided into two sub-rectangles with heights a_1 and a_2 and the same base b as in Fig. 1.1 (i). Then the area of subrectangles will be $f(a_1, b)$ and $f(a_2, b)$ and the area of the full rectangle is $f(a_1 + a_2, b)$. We have

$$f(a_1 + a_2, b) = f(a_1, b) + f(a_2, b). \tag{1.2}$$

In a similar manner, we divide the rectangle vertically with bases b_1 and b_2 and the same height a as in Fig. 1.1 (ii). Then the resulting areas are $f(a, b_1)$ and $f(a, b_2)$ and $f(a, b_1 + b_2)$. Therefore

$$f(a, b_1 + b_2) = f(a, b_1) + f(a, b_2). \tag{1.3}$$

In Eq. (1.1), b is a constant and in Eq. (1.3), a is a constant. Both the equations are similar to Cauchy's equation $f(x + y) = f(x) + f(y)$ whose solution is $f(x) = cx$. Therefore the solution of (1.2) and (1.3) is

$$f(a, b) = c_1(b)a = c_2(a)b. \tag{1.4}$$

From (1.4)

$$\frac{c_1(b)}{b} = \frac{c_2(a)}{a} = c. \tag{1.5}$$

From (1.5)

$$c_1(b) = cb, \qquad c_2(a) = ca. \tag{1.6}$$

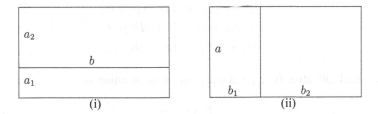

Fig. 1.1 Rectangles with base b and height a

Substitute (1.6) in (1.4), we get

$$f(a, b) = cab$$

where c is an arbitrary positive constant. Assume the initial conditions, that is, when $a = 1, b = 1$, the area of the rectangle $= 1$, which gives $c = 1$. Therefore, $f(a, b) = ab$. Hence we arrive the area of the rectangle.

1.3   Application of Functional Equation in Financial Management

Compound Interest: Suppose a person invests a principal of Rs.x at the rate of interest $r\%$ for the period of y years. One is interested in finding the formula for compound interest. The final amount is a function of x and y, that is, $f(x, y)$. There are two cases arise:

(i) The final amount is same, if we invest the principal amount $x_1 + x_2$ together or principal x_1, x_2 are invested separtely for y years. This can be expressed in the following functional equation

$$f(x_1 + x_2, y) = f(x_1, y) + f(x_2, y) \tag{1.7}$$

whose solution is

$$f(x, y) = D(y)x. \tag{1.8}$$

(ii) The final amount is same, whether we invest the amount x for a period of $y_1 + y_2$ years or x invested for y_1 years and then invest the resultant amount in y_2 years. It is expressed as

$$f(x, y_1 + y_2) = f(f(x, y_1), y_2). \tag{1.9}$$

Applying (1.8) in (1.9), we get

$$D(y_1 + y_2)x = f(D(y_1)x, y_2)$$
$$D(y_1 + y_2)x = D(y_2)D(y_1)x$$
$$D(y_1 + y_2) = D(y_2)D(y_1).$$

It is a multiplicative functional equation whose solution is

$$D(y) = a^y. \tag{1.10}$$

Substituting (1.10) in (1.8), we get

$$f(x, y) = a^y x.$$

For one year: $f(x, 1) = x +$ interest. Hence,

$$ax = x + \frac{x \times 1 \times r}{100}$$

$$ax = x \left(1 + \frac{r}{100}\right)$$

$$a = \left(1 + \frac{r}{100}\right).$$

Therefore, $f(x, y) = x \left(1 + \frac{r}{100}\right)^y.$

So, we arrive the compound interest formula $= x \left(1 + \frac{r}{100}\right)^y.$

1.4 Application of Functional Equation in Information Theory

Some functional equations like

(i) $f(x) + (1 - x)^\beta f \left(\frac{y}{1-x}\right) = f(y) + (1 - y)^\beta f \left(\frac{x}{1-y}\right)$

(ii) $f(xy) + f((1 - x)y) = f(y)\{m(x) + m(1 - x)\} + m(y)\{f(x) + f(1 - x)\}$

(iii) $\sum_{i=1}^{n} \sum_{j=1}^{m} f(x_i y_j) = \sum_{i=1}^{n} f(x_i) + \sum_{j=1}^{m} f(y_j)$

(iv) $\sum_{i=1}^{n} \sum_{j=1}^{m} f_{ij}(x_i y_j) = \sum_{i=1}^{n} g_i(x_i) + \sum_{j=1}^{m} h_j(y_j) + \sum_{i=1}^{n} k_i(x_i)$
      $\sum_{j=1}^{m} l_j(y_j)$

are applied in the information theory.

Definition 1.4.1 Let $\Delta_n = \{p = (p_1, p_2, \ldots, p_n) | p_i \geq 0, \sum_i p_i = 1\}$ be the set of all finite complete discrete probability distribution on a given partition of the sure event Ω into n events $E_1, E_2, \ldots, E_n$. In 1948, Shannon, in his paper [*C.E. Shannon, A Mathematical theory of Communication, Bell System Tech. J. 27 (1948), 378-423 and 632-656*] introduced the measure of information

$$H_n(p) = - \sum_{i=1}^{n} p_i log p_i, \qquad p \in \Delta_n$$

known as **Shannon's entropy**.

We have multiplicative functional equation

$$f(xy) = f(x) + f(y)$$

whose solution is $f(x) = log x$. We note that this is a function whose value on the product of probabilities of events is equal to the sum of its values on the probabilities of the individual events. Shannon used the above functional equation, in the information theory since with an intuitive notion that the information content of two independent events should be the sum of the information in each.

In particular, the functional equation

$$f(x) + \alpha(1-x)g\left(\frac{y}{1-x}\right) = h(y) + \alpha(1-y)k\left(\frac{x}{1-y}\right)$$

for all $x, y \in [0, 1]$ with $x + y \in [0, 1]$. When $f = g = h = k$ and $\alpha = $ the identity map is known as the fundamental equation of information. It has been extensively investigated by many authors. The general solution of the above equation is dealt by P.L. Kannappan [*Can. J. Math. Vol. XXXV, No. 5, 1983, pp. 862-872*].

The effect of the information communicated in a message can be measured by the changes in the probability concerning the receiver of the message. The effect of information will depend upon the expectation of receiver before and after receiving the message. Naturally the information received can be taken as the ratio of the logarithm of two probabilities. Thus the information received about the event E is given by

$$I(E) = \frac{\text{Probability concerning the receiver after receiving the information}}{\text{Probability concerning the receiver before receiving the information}}. \tag{1.11}$$

In case of noiseless channel, probability concerning the receiver after receiving the information equals to 1, as there will be no distortion of information during the process. The above equation becomes

$$I(E)$$
$$= \log\left[\frac{1}{\text{Probability concerning the receiver before receiving the information}}\right]$$
$$= -\log[\text{Probability concerning the receiver before receiving the information}]. \tag{1.12}$$

Shannon, with his intuitive idea, proposed a decreasing function $h(p)$, as a measure of the amount of information satisfying

$$h(p) = -\log p, \qquad 0 < p \le 1. \tag{1.13}$$

The function $h(p)$ is called the information function and it satisfies the additive property. Let A and B be any two events with $p(A) > 0$, $p(B) > 0$. Suppose that first we are informed that A has occurred and next that we are informed that B has occured and if A and B are independent, then

$$-\log[p(A)] - \log[p(B)] = -\log[p(AB)].$$

If $p(A) = p_1$, $p(B) = p_2$ and $p(AB) = p_1 p_2$, then

$$-\log(p_1) - \log(p_2) = -\log(p_1 p_2).$$

Hence $h(p_1) + h(p_2) = h(p_1 p_2)$.

It shows that the information function h satisfies the Cauchy's functional equation

$$f(xy) = f(x) + f(y). \tag{1.14}$$

1.5 Application of Functional Equation in Wireless Sensor Networks

Wireless Sensor Networks (WSNs) consist of small nodes with sensing, computation, and wireless communications capabilities. Many routing, power management, and data dissemination protocols have been specially designed for WSNs where energy awareness is an essential design issue.

Due to recent technological advances, the manufacturing of small and low cost sensors became technically and economically feasible. The sensing electronics measure ambient conditions related to the environment surrounding the sensor and transforms them into an electric signal. Processing such a signal reveals some properties about objects located and/or events happening in the vicinity of the sensor. A large number of these disposable sensors can be networked in many applications that require unattended operations. A Wireless Sensor Network (WSN) contains hundreds or thousands of these sensor nodes. These sensors have the ability to communicate either among each other or directly to an external base-station (BS). A greater number of sensors allows for sensing over larger geographical regions with greater accuracy. Basically, each sensor node comprises sensing, processing, transmission, mobilizer, position finding system, and power units (some of these components are optional like the mobilizer). Sensor nodes are usually scattered in a sensor field, which is an area where the sensor nodes are deployed. Sensor nodes coordinate among themselves to produce high-quality information about the physical environment. Each sensor node bases its decisions on its mission, the information it currently has, and its knowledge of its computing, communication, and energy resources. Each of these scattered sensor nodes has the capability to collect and route data either to other sensors or back to an external base station(s).

Networking unattended sensor nodes may have profound effect on the efficiency of many military and civil applications such as target field imaging, intrusion detection, weather monitoring, security and tactical surveillance, distributed computing, detecting ambient conditions such as temperature, movement, sound, light, or the presence of certain objects, inventory control, and disaster management. Deployment of a sensor network in these applications can be in random fashion (e.g., dropped from an airplane) or can be planted manually (e.g., fire alarm sensors in a facility). For example, in a disaster management application, a large number of sensors can be dropped from a helicopter. Networking these sensors can assist rescue operations by locating survivors, identifying risky areas, and making the rescue team more aware of the overall situation in the disaster area.

Routing is the process of selecting path in a network along which to send network traffic. Routing trees are typical structures used in WSN to deliver data to sink. To ensure robust data communication, efficient methods are required to choose routes across a network that can react quickly to communication link changes. Many algorithms have been proposed in literature to support the routing protocols of the network. In 1958, Richard Bellman [*Dynamic Programming, Princeton University Press, 1957*] applied the functional equation approach to devise an algorithm which converges to the solution at almost $N - 1$ steps for a network with N nodes.

It is stated as follows:

"Given a set of N cities, with every two cities linked by a road. The time required to travel from i to j is not directly proportional to the distance between i and j, due to road conditions and traffic. Given the matrix $T = (t_{ij})$ not necessarily symmetric, where t_{ij} is the time required to travel from i to j. We wish to determine the path from one given city to another given city which minimizes the travel time".

The functional equation technique of dynamic programming, combined with approximation in policy space, yields an iterative algorithm which converges after atmost $(N - 1)$ iterations.

Let us now introduce the functional equation technique of dynamic programming. Let $f_i =$ the time required to travel from i to N, $i = 1, 2, \ldots, N - 1$, using an optimal policy with $f_n = 0$.

Employing the principle of optimality, we see that the f_i satisfy the non-linear system of equations

$$f_i = \min[t_{ij} + f_i], \quad i = 1, 2, \ldots, N - 1 \tag{1.15}$$
$$f_N = 0.$$

The Eq. (1.15) is a functional equation because functions appear on both sides. We try to obtain the solution of the system (1.15) by using method of successive approximations. Choose an initial sequence $\left\{ f_i^{(0)} \right\}$, and then proceed iteratively, setting

$$f_i^{(k+1)} = \min_{i \neq j} \left(t_{ij} + f_i^{(k)} \right), \quad i = 1, 2, \ldots, N - 1 \tag{1.16}$$
$$f_N^{(k+1)} = 0, \text{ for } k = 0, 1, 2, \ldots$$

The sequence in (1.16) converges to the solution after $(N - 1)$ iterations by suitable algorithm. In this way, we can solve routing problem by functional equation.

1.6 Interpretation of Multiplicative Inverse Functional Equations

In this section, we deal with the following multiplicative inverse functional equation

$$r(x + y) = \frac{r(x)r(y)}{r(x) + r(y)}. \tag{1.17}$$

It is easy to see the function $r(x) = \frac{c}{x}$ is a solution of the Eq. (1.17). We call the Eq. (1.17) as reciprocal functional equation [121].

1.6.1 Geometrical Interpretation of Eq. (1.17)

Consider a right-angled triangle ABC with 'a' and 'b' as sides shown in the Fig. 1.2.

Construct a square BDEF inside the triangle ABC as shown in Fig. 1.2 with side 'p'. Then $AD = a - p$, $FC = b - p$. Now,

$$\text{Area of triangle ABC} = \frac{1}{2}ab. \tag{1.18}$$

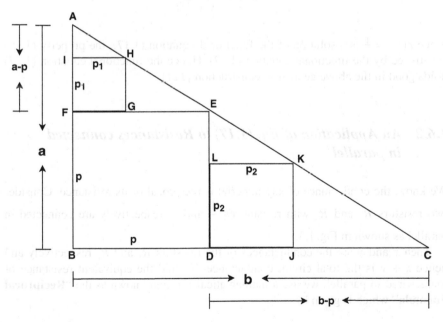

Fig. 1.2 Geometrical interpretation of Eq. (1.17)

From the Eq. (1.18), it is easy to show that $p = \frac{ab}{a+b}$. Now, construct another two squares FGHI and DJKL with sides 'p_1' and 'p_2' respectively as shown in the Fig. 1.2. Then $p_1 = \frac{p(a-p)}{a} = \frac{a^2b}{(a+b)^2}$ and $p_2 = \frac{p(b-p)}{b} = \frac{ab^2}{(a+b)^2}$. Now, $p_1 + p_2 = \frac{ab}{(a+b)^2}(a+b) = \frac{ab}{a+b} = p$. Therefore,

$$p_1 + p_2 = p. \tag{1.19}$$

In this construction, if we take $a = \frac{1}{x}, b = \frac{1}{y}$, then

$$p_1 = \frac{y}{(x+y)^2}, \tag{1.20}$$

$$p_2 = \frac{x}{(x+y)^2} \tag{1.21}$$

and

$$p = \frac{\frac{1}{x}\frac{1}{y}}{\frac{1}{x} + \frac{1}{y}} \tag{1.22}$$

Substituting the relations (1.20), (1.21) and (1.22) in (1.19), we get

$$\frac{1}{x+y} = \frac{\frac{1}{x}\frac{1}{y}}{\frac{1}{x} + \frac{1}{y}}.$$

Since $r(x) = \frac{1}{x}$ is a solution of the functional equation (1.17), the property (1.19) is satisfied by the functional equation (1.17). Hence the functional equation (1.17) holds good in the above geometric construction [121].

1.6.2   An Application of Eq. (1.17) to Resistances connected in parallel

We know, the conductance of any material is reciprocal of its resistance. Consider two resistors R_1 and R_2 with resistances $\frac{1}{x}$ and $\frac{1}{y}$ respectively are connected in parallel as shown in Fig. 1.3.

Then x and y are the conductances of the resistors R_1 and R_2 respectively and hence $x + y$ is the total circuit conductance. To find the equivalent resistance of loads wired in parallel, we use a mathematical formula known as the "**Reciprocal Formula**" which is given by

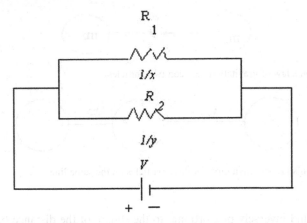

Fig. 1.3 Parallel circuits with couple of resistors

$$\frac{1}{\text{Total circuit conductance}} = \text{Total equivalent resistance of the parallel circuit.}$$

(1.23)

The reciprocal formula (1.23) satisfies the following algebraic identity

$$\frac{1}{x+y} = \frac{\frac{1}{x}\frac{1}{y}}{\frac{1}{x}+\frac{1}{y}}$$

and hence the functional equation (1.17) holds good in the above circuit [121].

1.7 Relevance of RQD and RQA Functional Equations

In this section, we focus on the following functional equation in two variables

$$f\left(\frac{x_1+x_2}{2}\right) - f(x_1+x_2) = \frac{3f(x_1)f(x_2)}{f(x_1)+f(x_2)+2\sqrt{f(x_1)f(x_2)}}.$$

(1.24)

and

$$f\left(\frac{x_1+x_2}{2}\right) + f(x_1+x_2) = \frac{5f(x_1)f(x_2)}{f(x_1)+f(x_2)+2\sqrt{f(x_1)f(x_2)}}.$$

(1.25)

We establish the geometrical descriptions of functional equations (1.30) and (1.31) using Newton's law of gravitation.

Newton's law of universal gravitation states that a particle attracts every other particle in the universe with a force which is directly proportional to the product of

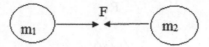

Fig. 1.4 Newton's law of gravitation between two particles

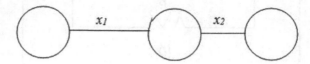

Fig. 1.5 Newton's law of gravitation for three particles on the same line

their masses and inversely proportional to the square of the distance between their centres (Figs. 1.4 and 1.5).

That is,

$$F = G\frac{m_1 m_2}{r^2}$$

where F is the force between the masses; G is the gravitational constant; m_1 is the first mass; m_2 is the second mass and r is the distance between the centres of the masses.

Suppose both the above two objects are of unit mass, then the force of attraction between them is

$$F = \frac{G}{x^2}$$

where x is the distance between the objects.

Consider three objects of unit mass with the following situation. Let x_1 be the distance between object 1 and object 2. Let x_2 be the distance between object 2 and object 3.

The force of attraction between object 1 and object 2 is

$$F(x_1) = \frac{G}{x_1^2}. \tag{1.26}$$

The force of attraction between object 2 and object 3 is

$$F(x_2) = \frac{G}{x_2^2}. \tag{1.27}$$

The force of attraction between object 1 and object 3 is

$$F(x_1 + x_2) = \frac{G}{(x_1 + x_2)^2}. \tag{1.28}$$

Using (1.26), (1.27) and (1.28), we have a relation as follows:

$$F(x_1 + x_2) = \frac{G}{x_1^2 + x_2^2 + 2x_1x_2}$$

$$= \frac{G}{x_1^2 x_2^2 \left(\frac{1}{x_1^2} + \frac{1}{x_2^2} + \frac{2}{x_1 x_2}\right)}$$

$$= \frac{\frac{G}{x_1^2 x_2^2}}{\frac{1}{x_1^2} + \frac{1}{x_2^2} + \frac{2}{x_1 x_2}}$$

$$= \frac{F(x_1)F(x_2)}{F(x_1) + F(x_2) + 2\sqrt{F(x_1)F(x_2)}}. \tag{1.29}$$

It is not hard to verify that the reciprocal-quadratic function $F(x) = \dfrac{c}{x^2}$, where c is a constant, is a solution of the above Eq. (1.29). Thus the reciprocal-quadratic functional equation (1.29) arises in the above physical phenomena.

If the distance between object 1 and object 3 is halved, that is, $\dfrac{x_1 + x_2}{2}$, then the force of attraction between them becomes,

$$F\left(\frac{x_1 + x_2}{2}\right) = \frac{G}{\left(\frac{x_1 + x_2}{2}\right)^2} = 4\frac{G}{(x_1 + x_2)^2}.$$

The relation between the difference of the forces $F\left(\dfrac{x_1 + x_2}{2}\right)$ and $F(x_1 + x_2)$ and the forces between object 1 and object 2; and object 2 and object 3 can be modeled as a functional equation as follows:

$$F\left(\frac{x_1 + x_2}{2}\right) - F(x_1 + x_2) = \frac{3G}{(x_1 + x_2)^2}$$

$$= \frac{3F(x_1)F(x_2)}{F(x_1) + F(x_2) + 2\sqrt{F(x_1)F(x_2)}}. \tag{1.30}$$

We call the above functional equation (1.30) as Reciprocal-Quadratic Difference Functional Equation (RQDF Equation) in two variables x_1 and x_2.

Similarly, the relation between the sum of the forces $F\left(\dfrac{x_1 + x_2}{2}\right)$ and $F(x_1 + x_2)$ and the forces between object 1 and object 2; and object 2 and object 3 can be modeled as a functional equation as follows:

$$F\left(\frac{x_1 + x_2}{2}\right) + F(x_1 + x_2) = \frac{5G}{(x_1 + x_2)^2}$$

$$= \frac{5F(x_1)F(x_2)}{F(x_1) + F(x_2) + 2\sqrt{F(x_1)F(x_2)}}. \tag{1.31}$$

The above functional equation (1.31) is said to be Reciprocal-Quadratic Adjoint Functional Equation (RQAF Equation) with a couple of variables x_1 and x_2.

1.8   Ulam's Motivation Problem for Stability of Functional Equation

The problem related to the stability of functional equations is instigated by the question of Ulam. Mathematically, the meaning of stability implies that it is a condition such that a negligible disturbance in a system does not affect much on that system. If there is an approximate solution which is near to the exact solution of an equation exists, then we say that the equation is stable. Since there are many physical applications of modeling involve deviations which certainly result from errors in measurement, the stability of equation is significant. Inspite of such differences, a stable solution is realistic.

Ulam [142] was the first mathematician to pose the primary query regarding stability of functional equations in 1940. His query paved a basis for the study of stability of functional equations. His query connected with the stability of homomorphisms between a metric group and a group, which is presented below:

"Consider a group $\mathcal{A}_1$. Let $d(\cdot, \cdot)$ be the metric defined on a group $\mathcal{A}_2$. Let ϵ be a positive constant. Then is it possible to persist a positive constant ρ with a mapping $h : \mathcal{A}_1 \longrightarrow \mathcal{A}_2$ fulfilling the condition $d(h(uv), h(u)u(v)) < \rho$ for all $u, v \in \mathcal{A}_1$ and a homomorphism $g : \mathcal{A}_1 \longrightarrow \mathcal{A}_2$ with $d(h(u), g(u)) < \epsilon$ for all $u \in \mathcal{A}_1$?"

If there is a response to the above query, then the functional equation for homomorphisms is said to be stable. This query is the beginning of investigation of stability of functional equation.

1.9   Stability of Functional Equation Incorporated with a Positive Constant (or H-U Stability)

In 1941, Hyers [57] was the first mathematician to present the first valuable contribution to Ulam's question concerning the stability of functional equations. He brilliantly proved the following celebrated theorem.

Theorem 1.9.1 ([57]) *Suppose $\mathcal{X}$ and $\mathcal{Y}$ are Banach spaces. Let $\phi : \mathcal{X} \longrightarrow \mathcal{Y}$ be a mapping satisfying*

$$\|\phi(u + v) - \phi(u) - \phi(v)\| \leq \epsilon \tag{1.32}$$

for all $u, v \in \mathcal{X}$. Then we can define an inimitable additive mapping $\mathcal{A} : \mathcal{X} \longrightarrow \mathcal{Y}$ which fulfills

$$\|\phi(u) - \mathcal{A}(u)\| \leq \epsilon \tag{1.33}$$

for all $u \in \mathcal{X}$ and the mapping $\mathcal{A}$ is defined by the existence of the limit

$$\mathcal{A}(u) = \lim_{m \to \infty} \frac{\phi(2^m u)}{2^m} \tag{1.34}$$

for all $u \in \mathcal{X}$.

In lieu of the above proof of Hyers, **Hyers-Ulam stability** for the additive Cauchy equation $\phi(u + v) = \phi(u) + \phi(v)$ is valid on $(\mathcal{X}, \mathcal{Y})$. The method provided by Hyers is called as a direct method. The works of Hyers is still decesive and initiated much of the present days research in stability theory of functional equations.

1.10   Stability of Functional Equation Incorporated with Sum of Exponents of Norms (or H-U-R Stability)

In 1951, Aoki [4] further taken broad view of Hyers theorem for approximately linear transformation in Banach spaces. The outcome obtained by Aoki is called **Hyers-Ulam-Aoki stability**. In 1978, Rassias [114] further generalized the Hyers' result by diminishing the condition for the Cauchy disparity and pioneered in substantiating which is now known to be **Hyers-Ulam-Rassias stability or H-U-R Stability**. This jargon is validated since the result of T. M. Rassias has strongly persuaded mathematicians who are involved in the research of stability problems of functional equations. T. M. Rassias proved the following theorem for sum of exponents of norms.

Theorem 1.10.1 ([114]) *Presume that $\mathcal{M}$ and $\mathcal{N}$ be Banach spaces. Suppose $0 \leq \theta < \infty$ and $0 \leq \beta < 1$. If $\phi : \mathcal{M} \longrightarrow \mathcal{N}$ is a mapping fulfills the condition*

$$\|\phi(u + v) - \phi(u) - \phi(v)\| \leq \theta(\|u\|^{\beta} + \|v\|^{\beta}) \tag{1.35}$$

for all $u, v \in \mathcal{M}$, then an inimitable additive function $T : \mathcal{M} \longrightarrow \mathcal{N}$ persists such that

$$\|\phi(u) - T(u)\| \leq \frac{2\theta}{2 - 2^{\beta}} \|u\|^{\beta} \tag{1.36}$$

for all $u \in \mathcal{M}$. Additionally, if $\phi(su)$ is continuous in s for each fixed $s \in \mathcal{M}$, then the function T is additive.

Later, Theorem 1.10.1 of T. M. Rassias was broadened for all $\alpha \neq 1$. T. M. Rassias observed that the theorem is still valid for $\alpha < 0$. In the interim of 27th International Conference on Functional Equations (1989), he asked whether his result can be proved for $\alpha \geq 1$. In 1991, Gajda [36] proved that the result was still true for $\alpha > 1$ by replaying m by $-m$ in the formula (1.34) and he also illustrated that the result was false when $\alpha = 1$ through the following counter example.

For a fixed $\lambda > 0$ and $\mu = \frac{1}{6}\lambda$, define a function $f : \mathbb{R} \longrightarrow \mathbb{R}$ by

$$f(u) = \sum_{m=0}^{\infty} 2^{-m} \phi(2^m u)$$

where the function $\phi : \mathbb{R} \longrightarrow \mathbb{R}$ given by

$$\phi(u) = \begin{cases} \mu & \text{for } u \in [1, \infty), \\ \mu u & \text{for } u \in (-1, 1), \\ -\mu & \text{for } u \in (-\infty, -1]. \end{cases}$$

Then the function f serves as a counter example for $\alpha = 1$ as presented in the following theorem.

Theorem 1.10.2 ([36]) *The function f defined above satisfies*

$$|f(u + v) - f(u) - f(v)| \leq \lambda(|u| + |v|) \tag{1.37}$$

for all $u, v \in \mathbb{R}$, while there is no constant $\Delta \geq 0$ and no additive function $A : \mathbb{R} \longrightarrow \mathbb{R}$ satisfying the condition

$$|f(u) - A(u)| \leq \Delta|u| \tag{1.38}$$

for all $u \in \mathbb{R}$.

From the above theorem of T. M. Rassias and the illustration provided by Gajda, $\alpha = 1$ is the only singular case for which the stability result fails.

1.11 Stability of Functional Equation Incorporated with Product of Exponents of Norms (or U-G-R Stability)

In 1982, John Micahel Rassias [110] presented an additional development further to the result of Hyers by proving the following using product of exponents of norms as upper bound. His theorem is in the sequel.

Theorem 1.11.1 ([110]) *Suppose $\mathcal{A}$ is a normed linear space with norm $\|.\|_1$. Let $\mathcal{B}$ be a Banach space with norm $\|.\|_2$. Also, let us assume in addition that $f : \mathcal{A} \longrightarrow \mathcal{B}$ is a mapping with the condition that $f(ru)$ is continuous in r for each fixed r. If there exist $s : 0 \leq s < \frac{1}{2}$ and $\epsilon \geq 0$ such that*

$$\|f(u + v) - [f(u) + f(v)]\|_2 \leq \epsilon\|u\|_1^s\|v\|_1^s \tag{1.39}$$

for any $u, v \in \mathcal{A}$, then there persists a unique linear mapping $\mathcal{L} : \mathcal{A} \longrightarrow \mathcal{B}$ such that

$$\|f(u) - \mathcal{L}(u)\|_2 \le C\|u\|_1^{2s} \tag{1.40}$$

for any $u \in \mathcal{A}$, where $C = \dfrac{\epsilon}{1 - 2^{2s-1}}$.

The outcome established by John Michal Rassias in the above theorem is termed as **Ulam-Gavruta-Rassias stability** or **U-G-R stability**.

1.12 Stability of Functional Equation Involving a General Control Function (or Generalized H-U-R Stability)

In 1994, Gavruta [38] modified Ulam's stability problem by swapping the upper bounds by a common control function. His result is given in the ensuing theorem.

Theorem 1.12.1 ([38]) *Let $(A, +)$ and $(X, \|\cdot\|)$ be an abelian group and be a Banach space, respectively. Let a function $\phi : A \times A \longrightarrow [0, \infty)$ fulfills*

$$\Phi(u, v) = \sum_{p=0}^{\infty} 2^{-p}\phi(2^p u, 2^p v) < \infty \tag{1.41}$$

for all $u, v \in A$. Let a function $f : A \longrightarrow X$ satisfies the condition

$$\|f(u + v) - f(u) - f(v)\| \le \phi(u, v) \tag{1.42}$$

for all $u, v \in G$. Then an inimitable function $T : A \longrightarrow X$ persists which fulfills the Cauchy additive equation and

$$\|f(u) - T(u)\| \le \frac{1}{2}\Phi(u, u) \tag{1.43}$$

for all $u \in A$.

The above result established by Găvruta is celeberated as **Generalized Hyers-Ulam-Rassias stability** or **Generalized H-U-R stability** of functional equation.

1.13 Stability Involving Assorted Type of Multiplication-Addition of Exponents of Norms (or JMR Stability)

The stability of the following quadratic functional equation

$$q(ku + v) + q(ku - v) = 2q(u + v) + 2q(u - v) + 2(k^2 - 2)q(u) - 2q(v)$$

is discussed by Ravi et al. [116], where k is an arbitrary constant but fixed real with the conditions $k \neq 0$; $k \neq \pm 1$; $k \neq \pm\sqrt{2}$ by involving assorted multiplication-addition of exponents of norms in the ensuing theorem.

Theorem 1.13.1 ([116]) *Let a function* $q : \mathcal{M} \longrightarrow \mathcal{F}$ *fulfills the condition*

$$\|q(ku + v) + q(ku - v) - 2q(u + v) - 2q(u - v) - 2(k^2 - 2)q(u) + 2q(v)\|_F$$
$$\leq \lambda \left\{ \|u\|_M^r \|v\|_M^r + \left(\|u\|_M^{2r} + \|v\|_M^{2r} \right) \right\} \tag{1.44}$$

for all $u, v \in \mathcal{M}$ *with* $u \perp v$, *where* λ *and* r *are constants with* $\lambda, r > 0$ *and either* $k > 1$; $r < 1$ *or* $k < 1$; $r > 1$ *with* $k \neq 0$; $k \neq \pm 1$; $k \neq \pm\sqrt{2}$ *and* $-1 \neq |k|^{r-1} < 1$. *Then the limit*

$$Q(u) = \lim_{p \to \infty} \frac{f(k^p u)}{k^{2p}} \tag{1.45}$$

exists for all $u \in \mathcal{M}$ *and also a distinctive quadratic function* $Q : \mathcal{M} \longrightarrow \mathcal{F}$ *persists which satisfies the condition of orthogonality with*

$$\|q(u) - Q(u)\|_F \leq \frac{\lambda}{2|k^2 - k^{2r}|} \|u\|_M^{2p}$$

for all $u \in \mathcal{M}$.

The afore stated proof is named as **J. M. Rassias stability involved with assorted product-sum of exponents of norms** by Ravi et al. [117–119, 122].

1.14 Various Types of Functional Equations

There are many interesting, motivating and novel results concerning the stability of various forms of several functional equations accomplished by many mathematicians which creat innovative thinking and decisive disputes.

The functional equation

$$\phi(u + v) = \phi(u) + \phi(v) \tag{1.46}$$

is the most celeberated equation among the functional equations. To honor Cauchy, this Eq. (1.46) is called additive Cauchy functional equation. The function $\phi(u) = cu$ is the solution of the functional equation (1.46). For developing various theories and properties of other type of functional equations, this equation (1.46) is applied often.

The functional equation

$$Q(u + v) + Q(u - v) = 2Q(u) + 2Q(v) \tag{1.47}$$

is called as a quadratic functional equation since its solution is the quadratic function $Q(u) = cu^2$. Skof [138] achieved the stability of the quadratic function equation (1.47) in the sense of Hyers and Ulam. There are also many interesting results concering various quadratic functional equations, one can refer [1, 8, 9, 19, 21, 22, 55, 76, 80, 109, 111].

Motivated by the quadratic functional equation (1.47), in 2001, Rassias [112] brought a new equation in the theory of functional equations

$$C(u + 2v) - 3C(u + v) + 3C(u) - C(u - v) = 6C(v) \qquad (1.48)$$

and achieved its general solution and investigated its Ulam stability. It is not hard to show that the function $C(u) = ku^3$ satisfies the functional equation (1.48) and hence it is called as cubic functional equation. Several other types of cubic functional equations and its stability problems were discussed by many authors, one can refer to [31, 51, 60, 62, 63, 82, 90, 100, 120, 144].

Another break through in the theory of functional equations was made by John Michael Rassias. He introduced the ensuing equation

$$q(u + 2v) + q(u - 2v) = 4q(u + v) + 4q(u - v) + 6q(u) + 24q(v). \qquad (1.49)$$

It can be verified that the function $q(u) = ku^4$ is a solution of the Eq. (1.49) and hence it is known as quartic functional equation.

The stability problems of several types of functional equations which include mixed type functional equations such as additive-quadratic, quadratic-cubic, additive-quadratic-cubic, were dealt by many mathematicians recently and accomplished their validity of Ulam stability and the stabilities associated with Rassias and Gavruta. For further details one can see [5, 18, 39–50, 61, 73, 77, 81, 91].

In recent times, Xu et al. [149] solved the Ulam stability problem for the general mixed type functional equation originating from additive, quadratic, cubic and quartic functions

$$h(u + nv) + h(u - nv) = n^2[h(u + v) + h(u - v)] + 2(1 - n^2)h(u)$$
$$+ \frac{n^4 - n^2}{12}[h(2v) + hf(-2v) - 4h(v) - 4h(-v)]$$

$$(1.50)$$

with $n \neq 0$ is a fixed integer, in multi-Banach spaces using fixed point technique.

The validity of Ulam stability was obtained by Park [95] for the ensuing functional equation developed from additive, quadratic, cubic and quartic functions

$$p(u + 2v) + p(u - 2v) = 4p(u + v) + 4p(u - v) - 4p(v) - 4p(-v)$$
$$+ p(2v) + p(-2v) - 6p(u)$$

in non-Archimedean normed spaces.

For the last 40 years, there are many research articles published in the theory of Ulam stability of functional equations for a numerous functional equations in various spaces, for detailed study one can refer [2, 3, 6, 7, 11–15, 17, 26, 32, 53, 71, 83, 92–94, 98, 101, 102, 106, 108, 121, 123, 132–135, 139, 141].

There are also many research monographs available in the hypothesis of stability of functional equations, one can refer to [23, 24, 58, 65, 67, 72, 113, 115].

Chapter 2
Stability and Instability of Multiplicative Inverse Type Tredecic and Quottuordecic Functional Equations in Non-archimedean Spaces

Abstract This chapter deals with the investigation of validity of various fundamental stabilities of multiplicative inverse type tredecic and quottuordecic functional equations relevant to Ulam stability theory in non-Archimedean fields via fixed point method. Two suitable counter-examples are also included to prove that the stability results are not valid for singular cases.

2.1 Introduction

The p-adic numbers were discovered by Hensel [54] in 1897 as a number theoretical analogue of power series in complex analysis. He introduced a field with a valuation norm which does not satisfy the Archimedean property.

In this chapter, we introduce a multiplicative inverse tredecic functional equation

$$m_t(2u + v) + m_t(2u - v)$$
$$= \frac{4m_t(u)m_t(v)}{\left(4m_t(v)^{2/13} - m_t(u)^{2/13}\right)^{13}} \left[\frac{1}{2}\sum_{k=0}^{6}\binom{13}{2k}[m_t(u)]^{2k/13}[m_t(v)]^{(13-2k)/13}\right]$$

$$(2.1)$$

and a multiplicative inverse quottuordecic functional equation.

$$m_q(2u + v) + m_q(2u - v)$$
$$= \frac{4m_q(u)m_q(v)}{\left(4m_q(v)^{1/7} - m_q(u)^{1/7}\right)^{14}} \left[\sum_{k=0}^{7}\binom{14}{2k}[m_q(u)]^{k/7}[m_q(v)]^{(14-2k)/14}\right].$$

$$(2.2)$$

B. V. Senthil Kumar and H. Dutta, *Multiplicative Inverse Functional Equations*,
Studies in Systems, Decision and Control 289, https://doi.org/10.1007/978-3-030-45355-8_2

It is easy to verify that the multiplicative inverse tredecic function $m_t(u) = \dfrac{1}{u^{13}}$
and the multiplicative inverse quottuordecic function $m_q(u) = \dfrac{1}{u^{14}}$ are solutions of
the Eqs. (2.1) and (2.2), respectively. By employing fixed point technique, we prove
various Ulam stability problems concerning the above Eqs. (2.1) and (2.2) in non-
Archimedean fields and also illustrate suitable counter-examples to invalidate the
stability results for singular cases.

2.2 Preliminaries

We introduce a few fundamental notions associated with non-Archimedean fields.
We also present the fixed point alternative principle in non-Archimedean version.

Definition 2.2.1 Suppose $\mathbb{F}$ is a field with a function (valuation) $|\cdot|$ from $\mathbb{F}$ into
$[0, \infty)$. Then $\mathbb{F}$ is called a non-Archimedean field if the following conditions hold:
(i) $|m| = 0$ if and only if $m = 0$; (ii) $|mn| = |m||n|$; (iii) $|m + n| \leq \max\{|m|, |n|\}$
for all $m, n \in \mathbb{F}$.

Obviously $|1| = |-1| = 1$ and $|m| \leq 1$ for all $m \in \mathbb{N}$. Furthermore, we presume
that $|\cdot|$ is non-trivial, that is, there exists an $\alpha_0 \in \mathbb{F}$ such that $|\alpha_0| \neq 0, 1$.

Let X be a vector space over a scalar field $\mathbb{K}$ with a non-Archimedean non-
trivial valuation $|\cdot|$. A function $\|\cdot\| : X \longrightarrow \mathbb{R}$ is a *non-Archimedean norm (valuation)*
if it satisfies the following conditions: (i) $\|u\| = 0$ if and only if $u = 0$;
(ii) $\|\rho u\| = |\rho|\|u\|$ ($\rho \in \mathbb{K}, u \in X$); (iii) the strong triangle inequality (ultramet-
ric); namely,

$$\|u + v\| \leq \max\{\|u\|, \|v\|\} \qquad (u, v \in X).$$

Then $(X, \|\cdot\|)$ is called a non-Archimedean space. By virtue of the inequality

$$\|u_n - u_m\| \leq \max\left\{\|u_{j+1} - u_j\| : m \leq j \leq n - 1\right\} \qquad (n > m),$$

a sequence $\{u_n\}$ is Cauchy if and only if $\{u_{n+1} - u_n\}$ converges to zero in a non-
Archimedean space. By a complete non-Archimedean space, we mean that every
Cauchy sequence is convergent in the space.

An example of a non-Archimedean valuation is the mapping $|\cdot|$ taking everything
but 0 into 1 and $|0| = 0$. This valuation is called trivial. Another example of a non-
Archimedean valuation on a field $\mathbb{K}$ is the mapping

$$|\tau| = \begin{cases} 0 & \text{if } \beta = 0 \\ \frac{1}{\beta} & \text{if } \beta > 0 \\ -\frac{1}{\beta} & \text{if } \beta < 0 \end{cases}$$

for any $\beta \in \mathbb{K}$.

Example 2.2.2 For a given prime number p, the p-adic absolute value in $\mathbb{Q}$ is defined as follows: If u is a non-zero rational number, then there exists a unique integer r such that $u = p^r \frac{a}{b}$, where a and b are coprime to p. Set $ord_p(u) = r$ and $|u|_p = p^{-r}$. Then $ord_p(u)$ is called the p-adic valuation of u and $|u|_p$ is called the p-adic absolute value of u.

  (i) By the definition of $|\cdot|_p$, it is clear that $|u|_p = 0$ if and only if $u = 0$.
 (ii) $|u|_p |v|_p = p^{-ord_p(u)} p^{-ord_p(v)}$. By the fundamental theorem of arithmetic, the number of prime factors p in uv is the same as the sum of the factors in u and v individually $(ord_p(u) + ord_p(v))$. Hence, we have $|uv|_p = p^{-ord_p(u)} p^{-ord_p(v)}$.
(iii) $|u + v|_p = p^{-ord_p(u+v)} \leq max\left\{p^{-ord_p(u)}, p^{-ord_p(v)}\right\} = max\left\{|u|_p, |v|_p\right\}$.

Hence, the p-adic absolute value defined above is a non-Archimedean norm on $\mathbb{Q}$. The completion of $\mathbb{Q}$ with respect to $|\cdot|$ which is denoted by $\mathbb{Q}_p$ is said to be the p-adic number field. Note that if $p > 2$, then $|2^n| = 1$ for all integers n.

Definition 2.2.3 Let A be a nonempty set and $d : A \times A \to [0, \infty]$ satisfy the ensuing properties: (i) $d(a, b) = 0$ if and only if $a = b$; (ii) $d(a, b) = d(b, a)$ (symmetry); (iii) $d(a, c) \leq max\{d(a, b), d(b, c)\}$ (strong triangle inequality) for all $a, b, c \in A$. Then (A, d) is called a generalized non-Archimedean metric space. (A, d) is called complete if every Cauchy sequence in A is convergent.

Example 2.2.4 For each nonempty set A, define

$$d(u, u^\star) = \begin{cases} 0 & \text{if } u = u^\star \\ \infty & \text{if } u \neq u^\star. \end{cases}$$

Then d is a generalized non-Archimedean metric on A.

Example 2.2.5 Let A and B be two non-Archimedean spaces over a non-Archimedean field $\mathbb{K}$. If B has a complete non-Archimedean norm over $\mathbb{K}$ and $\phi : A \longrightarrow [0, \infty)$, for each $s, t : A \longrightarrow B$, define

$$d(s, t) = \inf\{\delta > 0 : |s(u) - t(u)| \leq \delta\phi(u), \ \forall u \in A\}.$$

Using Theorem 2.5 [17], Mirmostafaee [76] introduced non-Archimedean version of the alternative fixed point theorem as follows:

Theorem 2.2.6 [76] (Non-Archimedean Alternative Contraction Principle) *If (A, d) is a non-Archimedean generalized complete metric space and $\Gamma : A \to A$ a strictly contractive mapping (that is $d(\Gamma(x), \Gamma(y)) \leq Ld(y, x)$, for all $x, y \in A$ and a Lipschitz constant $L < 1$), then either*

(i) $d\left(\Gamma^n(x), \Gamma^{n+1}x\right) = \infty$ for all $n \geq 0$, or
(ii) there exists some $n_0 \geq 0$ such that $d\left(\Gamma^n(x), \Gamma^{n+1}(x)\right) < \infty$ for all $n \geq n_0$;

the sequence $\{\Gamma^n(x)\}$ is convergent to a fixed point $x^\star$ of Γ; $x^\star$ is the unique fixed point of Γ in the set $Y = \{y \in X : d(\Gamma^{n_0}(x), y) < \infty\}$ and $d(y, x^\star) \leq d(y, \Gamma(y))$ for all y in this set.

Throughout this chapter, let us assume that $\mathbb{G}$ and $\mathbb{H}$ are a non-Archimedean field and a complete non-Archimedean field, respectively. We use the following notation: $\mathbb{G}^* = \mathbb{G}\backslash\{0\}$, where $\mathbb{G}$ is a non-Archimedean field. For the sake of simplification, we describe the difference operators $\Delta_1, \Delta_2 : \mathbb{G}^* \times \mathbb{G}^* \longrightarrow \mathbb{H}$ by

$$\Delta_1 m_t(u, v) = m_t(2u + v) + m_t(2u - v)$$
$$- \frac{4m_t(u)m_t(v)}{\left(4m_t(v)^{2/13} - m_t(u)^{2/13}\right)^{13}} \left[\frac{1}{2}\sum_{k=0}^{6} \binom{13}{2k} [m_t(u)]^{2k/13} [m_t(v)]^{(13-2k)/13}\right]$$

and

$$\Delta_2 m_q(u, v) = m_q(2u + v) + m_q(2u - v)$$
$$- \frac{4m_q(u)m_q(v)}{\left(4m_q(v)^{1/7} - m_q(u)^{1/7}\right)^{14}} \left[\sum_{k=0}^{7} \binom{14}{2k} [m_q(u)]^{k/7} [m_q(v)]^{(14-2k)/14}\right]$$

for all $u, v \in \mathbb{G}^*$.

2.3 Fundamental Stabilities of Eqs. (2.1) and (2.2)

In this section, we investigate fundamental stability results of Eqs. 2.1 and (2.2) in the framework of non-Archimedean fields, via fixed point method.

Definition 2.3.1 A mapping $m_t : \mathbb{G}^* \longrightarrow \mathbb{H}$ is called as multiplicative inverse tre-decic mapping if m_t satisfies the Eq. (2.1) and hence (2.1) is called as a multiplicative inverse tredecic functional equation. Also, a mapping $m_q : \mathbb{G}^* \longrightarrow \mathbb{H}$ is called as multiplicative inverse quottuordecic mapping if m_q satisfies the Eq. (2.2) and so (2.2) is said to be a multiplicative inverse quottuordecic functional equation.

Assumptions on the above definition and Eqs. (2.1) **and** (2.2): From the above definition, we find that the equalities $v = 2u$ and $v = -2u$ can not occur since $2u - v$ and $2u + v$ do not belong to $\mathbb{G}^*$. On the other hand, in (2.1) if $4m_t(v)^{2/13} - m_t(u)^{2/13} = 0$, then this is equivalent to $m_t(v) = m_t(2u)$. Since the multiplicative inverse tredecic function $m_t(u) = \frac{1}{u^{13}}$ is the solution of (2.1), this implies that $v = 2u$ which is impossible. But, by assuming $m_t(u) \neq 0, m_t(v) \neq 0, m_q(u) \neq 0, m_q(v) \neq 0, 4m_t(v)^{2/13} - m_t(u)^{2/13} \neq 0$ and $4m_q(v)^{1/7} - m_q(u)^{1/7} \neq 0$ for all $u, v \in \mathbb{G}^*$, we can avoid the singular cases.

Theorem 2.3.2 Let $k = \pm 1$ be a fixed number. Suppose a mapping $m_t : \mathbb{G}^* \longrightarrow \mathbb{H}$ satisfies the inequality

$$|\Delta_1 m_t(u, v)| \leq \zeta(s, t) \tag{2.3}$$

for all $u, v \in \mathbb{G}^*$, where $\zeta : \mathbb{G}^* \times \mathbb{G}^* \longrightarrow \mathbb{H}$ is a given function. If $0 < L < 1$,

$$|3|^{13k} \zeta\left(3^k u, 3^k v\right) \leq L\zeta(u, v) \tag{2.4}$$

for all $u, v \in \mathbb{G}^*$, then there exists a unique multiplicative inverse tredeic mapping $m_T : \mathbb{G}^* \longrightarrow \mathbb{H}$ satisfying the functional Eq. (2.1) and

$$|m_t(u) - m_T(u)| \leq L|3|^{13k}\zeta(u, u) \tag{2.5}$$

for all $u \in \mathbb{G}^*$.

Proof Let us consider the case $k = -1$. Now, plugging v into $\frac{u}{3}$ in (2.3), we obtain

$$\left| m_t(u) - \frac{1}{3^{13}}m_t\left(\frac{u}{3}\right) \right| \leq \zeta\left(\frac{u}{3}, \frac{u}{3}\right) \tag{2.6}$$

for all $u \in \mathbb{G}^*$. Let $\mathcal{F} = \{g | g : \mathbb{G}^* \longrightarrow \mathbb{H}\}$,

$$d(g, h) = \inf\{\beta > 0 : |g(u) - h(u)| \leq \beta\zeta(u, u), \quad \text{for all } u \in \mathbb{G}^*\}. \tag{2.7}$$

Now, let us prove that $(\mathcal{F}, d)$ is complete. Using the idea from [64], we prove the completeness of $(\mathcal{F}, d)$. Let $\{h_n\}$ be a Cauchy sequence in $(\mathcal{F}, d)$. Then for any $\epsilon > 0$, there exists an integer $N_\epsilon > 0$ such that $d(h_m, h_n) \leq \epsilon$ for all $m, n \geq N_\epsilon$. From (2.7), we arrive

$$\forall \epsilon > 0, \exists N_\epsilon \in \mathbb{N}, \forall m, n \geq N_\epsilon, \forall u \in \mathbb{G}^*, |h(u) - h_n(u)| \leq \epsilon\zeta(u, u). \tag{2.8}$$

If u is a fixed number, (2.8) implies that $\{h_n(u)\}$ is a Cauchy sequence in $(\mathbb{H}, |\cdot|)$. Since $(\mathbb{H}, |\cdot|)$ is complete, $\{h_n(u)\}$ converges for all $u \in \mathbb{G}^*$. Therefore, we can define a function $h : \mathbb{G}^* \longrightarrow \mathbb{H}$ by $h(u) = \lim_{n \to \infty} h_n(u)$ and hence $h \in \mathcal{F}$. Letting $m \to \infty$ in (2.8), we have

$$\forall \epsilon > 0, \exists N_\epsilon \in \mathbb{N}, \forall n \geq N_\epsilon, \forall u \in \mathbb{G}^* : |h(u) - h_n(u)| \leq \epsilon\zeta(u, u).$$

By considering (2.7), we arrive

$$\forall \epsilon > 0, \exists N_\epsilon \in \mathbb{N}, \forall n \geq N_\epsilon : d(h, h_n) \leq \epsilon,$$

which implies that the Cauchy sequence $\{h_n\}$ converges to h in $(\mathcal{F}, d)$. Hence $(\mathcal{F}, d)$ is complete.

Define $\rho : \mathcal{F} \longrightarrow \mathcal{F}$ by $\rho(g)(s) = 3^{-13}g\left(3^{-13}u\right)$ for all $s \in \mathbb{G}^*$ and $g \in \mathcal{F}$. Then ρ is strictly contractive on $\mathcal{F}$, in fact if $|g(u) - h(u)| \leq \beta\zeta(u, u)$, $(u \in \mathbb{G}^*)$, then by (2.4), we obtain

$$|\rho(g)(u) - \rho(h)(u)| = |3|^{-13}\left|g\left(3^{-1}u\right) - h\left(3^{-1}u\right)\right|$$
$$\leq \beta|3|^{-13}\zeta\left(3^{-1}u, 3^{-1}u\right) \leq \beta L\zeta(u, u) \quad (u \in \mathbb{G}^*).$$

From the above, we conclude that $(\rho(g), \rho(h)) \leq Ld(g, h)$ $(g, h \in \mathcal{F})$. Hence d is strictly contractive mapping with Lipschitz constant L. Using (2.6), we have

$$|\rho(m_t)(u) - m_t(u)| = \left|3^{-13}m_t\left(3^{-1}u\right) - m_t(u)\right| \leq \zeta\left(\frac{u}{3},\frac{u}{3}\right) \leq |3|^{13}L\zeta(u,u) \quad (u \in \mathbb{G}^*).$$

This indicates that $d(\rho(m_t), m_t) \leq L|3|^{13}$. Due to Theorem 2.2.6 (ii), ρ has a unique fixed point $m_T : \mathbb{G}^* \longrightarrow \mathbb{H}$ in the set $Y = \{g \in \mathcal{F} : d(m_t, g) < \infty\}$ and for each $u \in \mathbb{G}^*$,

$$m_T(u) = \lim_{n \to \infty} \rho^n m_t(u) = \lim_{n \to \infty} 3^{-13n} m_t\left(3^{-n}u\right) \quad (u \in \mathbb{G}^*).$$

Therefore, for all $u, v \in \mathbb{G}^*$,

$$\begin{aligned}
|\Delta_1 m_T(u, v)| &= \lim_{m \to \infty} |3|^{-13m} \left|\Delta_1 m_t\left(3^{-m}u, 3^{-m}v\right)\right| \\
&\leq \lim_{m \to \infty} |3|^{-13m} \zeta\left(3^{-m}u, 3^{-m}v\right) \\
&\leq \lim_{m \to \infty} L^m \zeta(u, v) = 0
\end{aligned}$$

which shows that m_T is multiplicative inverse tredecic mapping. By Theorem 2.2.6 (ii), we have $d(m_t, m_T) \leq d(\rho(m_t), m_t)$, that is, $|m_t(u) - m_T(u)| \leq |3|^{-13}$ $L\zeta(u, u), u \in \mathbb{G}^*$. Let $m'_T : \mathbb{G}^* \longrightarrow \mathbb{H}$ be a multiplicative inverse tredecic mapping which satisfies (2.5), then m'_T is a fixed point of ρ in $\mathcal{F}$. However, by Theorem 2.2.6, ρ has only one fixed point in Y. Similarly, the theorem can be proved for the case $k = 1$. □

The upcoming corollaries are direct consequences of Theorem 2.3.2. In the following corollaries, we assume that $|2| < 1$ for a non-Archimedean field $\mathbb{G}$.

Corollary 2.3.3 *Let δ (independent of u, v)≥ 0 be a constant exists for a mapping $m_t : \mathbb{G}^* \longrightarrow \mathbb{H}$ satisfies the inequality*

$$|\Delta_1 m_t(u, v)| \leq \delta$$

for all $u, v \in \mathbb{G}^*$. Then there exists a unique multiplicative inverse tredecic mapping $m_T : \mathbb{G}^* \longrightarrow \mathbb{H}$ satisfying Eq. (2.1) and

$$|m_t(u) - m_T(u)| \leq \delta$$

for all $u \in \mathbb{G}^*$.

Proof Assuming $\zeta(u, v) = \delta$ and selecting $L = |3|^{-13}$ in Theorem 2.3.2, we obtain the required result. □

Corollary 2.3.4 *Let $\beta \neq -13$ and $k_1 \geq 0$ be real numbers exist for a mapping $m_t : \mathbb{G}^* \longrightarrow \mathbb{H}$ such that*

$$|\Delta_1 m_t(u, v)| \leq k_1\left(|u|^\beta + |v|^\beta\right)$$

for all $u, v \in \mathbb{G}^*$. *Then there exists a unique multiplicative inverse tredecic mapping* $m_T : \mathbb{G}^* \longrightarrow \mathbb{H}$ *satisfying Eq. (2.1) and*

$$|m_t(u) - m_T(u)| \leq \begin{cases} \frac{|2|k_1}{|3|^\beta}|u|^\beta, & \beta > -13 \\ |2|k_1|3|^{13}|u|^\beta, & \beta < -13 \end{cases}$$

for all $u \in \mathbb{G}^*$.

Proof Considering $\zeta(u, v) = k_1 \left(|u|^\beta + |v|^\beta\right)$ in Theorem 2.3.2 and then assuming $L = |3|^{-\beta-13}, \beta > -13$ and $L = |3|^{\beta+13}, \beta < -13$, respectively, for each case of k, the proof follows directly. $\qquad\square$

Corollary 2.3.5 *Let* $k_2 \geq 0$ *and* $\beta \neq -13$ *be real numbers, and* $m_t : \mathbb{G}^* \longrightarrow \mathbb{H}$ *be a mapping satisfying the inequality*

$$|\Delta_1 m_t(u, v)| \leq k_2|u|^{\beta/2}|v|^{\beta/2}$$

for all $u, v \in \mathbb{G}^*$. *Then there exists a unique multiplicative inverse tredecic mapping* $m_T : \mathbb{G}^* \longrightarrow \mathbb{H}$ *satisfying Eq. (2.1) and*

$$|m_t(u) - m_T(u)| \leq \begin{cases} \frac{k_2}{|3|^\beta}|s|^\beta, & \beta > -13 \\ |3|^{13}k_2|s|^\beta, & \beta < -13 \end{cases}$$

for all $u \in \mathbb{G}^*$.

Proof It is easy to prove this corollary, by taking $\zeta(u, v) = k_2|u|^{\beta/2}|v|^{\beta/2}$ and then choosing $L = |3|^{-\beta-13}, \beta > -13$ and $L = |3|^{\beta+13}, \beta < -13$, respectively for each case k in Theorem 2.3.2. $\qquad\square$

We employ fixed point technique to investigate various stabilities of Eq. (2.2) in non-Archimedean fields. Since the proof of the following results are similar to the results obtained for the Eq. (2.1), for the sake of completeness, we provide only statement of theorems and skip their proofs.

Theorem 2.3.6 *Let* $k = \pm 1$. *Let* $m_q : \mathbb{G} \longrightarrow \mathbb{H}$ *be a mapping satisfying the inequality*

$$|\Delta_2 m_q(u, v)| \leq \eta(u, v)$$

for all $u, v \in \mathbb{G}^*$, *where* $\eta : \mathbb{G}^* \times \mathbb{G}^* \longrightarrow \mathbb{H}$ *is a given function. If* $0 < L < 1$,

$$|3|^{12k}\eta\left(3^k u, 3^k v\right) \leq L\eta(u, v)$$

for all $u, v \in \mathbb{G}^*$, then there exists a unique multiplicative inverse quottuordecic mapping $m_Q : \mathbb{G}^* \longrightarrow \mathbb{H}$ satisfying Eq. (2.2) and

$$|m_q(u) - m_Q(u)| \leq L|3|^{14k}\eta(u, u)$$

for all $u \in \mathbb{G}^*$.

Corollary 2.3.7 *Let ρ (independent of u, v)≥ 0 be a constant. Suppose a mapping $m_q : \mathbb{G}^* \longrightarrow \mathbb{H}$ satisfies the inequality*

$$\left|\Delta_2 m_q(u, v)\right| \leq \rho$$

for all $u, v \in \mathbb{G}^*$. Then there exists a unique multiplicative inverse quottuordecic mapping $m_Q : \mathbb{G}^* \longrightarrow \mathbb{H}$ satisfying Eq. (2.2) and

$$|m_q(u) - m_Q(u)| \leq \rho$$

for all $u \in \mathbb{G}^*$.

Corollary 2.3.8 *Let $\beta \neq -14$ and $\theta_1 \geq 0$ be real numbers. If $m_q : \mathbb{G}^* \longrightarrow \mathbb{H}$ is a mapping satisfies the inequality*

$$\left|\Delta_2 m_q(u, v)\right| \leq \theta_1 \left(|u|^\beta + |v|^\beta\right)$$

for all $u, v \in \mathbb{G}^*$, then there exists a unique multiplicative inverse quottuordecic mapping $m_Q : \mathbb{G}^* \longrightarrow \mathbb{H}$ satisfying Eq. (2.2) and

$$|m_q(u) - m_Q(u)| \leq \begin{cases} \frac{|2|\theta_1}{|3|^\beta}|u|^\beta, & \beta > -14 \\ |2|\theta_1|3|^{14}|u|^\beta, & \beta < -14 \end{cases}$$

for all $u \in \mathbb{G}^*$.

Corollary 2.3.9 *Let $m_q : \mathbb{G}^* \longrightarrow \mathbb{H}$ be a mapping and $\theta_2 \geq 0$ and $\beta \neq -14$ be real numbers. If the mapping m_q satisfies the inequality*

$$\left|\Delta_2 m_q(u, v)\right| \leq \theta_2|u|^{\beta/2}|v|^{\beta/2}$$

for all $u, v \in \mathbb{G}^*$, then there exists a unique multiplicative inverse quottuordecic mapping $m_Q : \mathbb{G}^* \longrightarrow \mathbb{H}$ satisfying Eq. (2.2) and

$$|m_q(u) - m_Q(u)| \leq \begin{cases} \frac{\theta_2}{|3|^\beta}|u|^\beta, & \beta > -14 \\ |3|^{14}\theta_2|u|^\beta, & \beta < -14 \end{cases}$$

for all $u \in \mathbb{G}^*$.

2.4 Counter-Examples

In this section, we show that the Eqs. (2.1) and (2.2) are not valid for $\beta = -13$ in Corollary 2.3.4 and $\beta = -14$ in Corollary 2.3.9, respectively, in the setting of non-zero real numbers.

Example 2.4.1 Let us define the function

$$\chi(u) = \begin{cases} \frac{c}{u^{13}}, & \text{for } u \in (1, \infty) \\ c, & \text{elsewhere} \end{cases} \tag{2.9}$$

where $\chi : \mathbb{R}^* \longrightarrow \mathbb{R}$. Let $m_t : \mathbb{R}^* \longrightarrow \mathbb{R}$ be a function defined as

$$m_t(u) = \sum_{m=0}^{\infty} 1594323^{-m} \chi(3^{-m} u) \tag{2.10}$$

for all $u \in \mathbb{R}$. Suppose the mapping $m_t : \mathbb{R}^* \longrightarrow \mathbb{R}$ described in (2.10) satisfies the functional inequality

$$|\Delta_1 m_t(u, v)| \leq \frac{2391485 \, c}{797161} \left(|u|^{-13} + |v|^{-13} \right) \tag{2.11}$$

for all $u, v \in \mathbb{R}^*$. We prove that there do not exist a multiplicative inverse tredecic mapping $m_T : \mathbb{R}^* \longrightarrow \mathbb{R}$ and a constant $\delta > 0$ such that

$$|m_t(u) - m_T(u)| \leq \delta \, |u|^{-13} \tag{2.12}$$

for all $u \in \mathbb{R}^*$. Firstly, let us prove that m_t satisfies (2.11). Using (2.9), we have

$$|m_t(u)| = \left| \sum_{m=0}^{\infty} 1594323^{-m} \chi(3^{-m} u) \right| \leq \sum_{m=0}^{\infty} \frac{c}{1594323^m} = \frac{1594323}{1594322} c.$$

We observe that m_t is bounded by $\frac{1594323 \, c}{1594322}$ on $\mathbb{R}$. If $|u|^{-13} + |v|^{-13} \geq 1$, then the left hand side of (2.11) is less than $\frac{2391485 \, c}{797161}$. Now, suppose that $0 < |u|^{-13} + |v|^{-13} < 1$. Hence, there exists a positive integer m such that

$$\frac{1}{1594323^{m+1}} \leq |u|^{-13} + |v|^{-13} < \frac{1}{1594323^m}. \tag{2.13}$$

Hence, the inequality (2.13) generates $1594323^m \left(|u|^{-13} + |v|^{-13} \right) < 1$, or equivalently; $1594323^m u^{-13} < 1$, $1594323^m v^{-13} < 1$. So, $\frac{u^{13}}{1594323^m} > 1$, $\frac{v^{13}}{1594323^m} > 1$. Hence, the last inequalities imply $\frac{u^{13}}{1594323^{m-1}} > 1594323 > 1$, $\frac{v^{13}}{1594323^{m-1}} >$

$1594323 > 1$ and as a result, we find $\frac{1}{3^{m-1}}(u) > 1$, $\frac{1}{3^{m-1}}(v) > 1$, $\frac{1}{3^{m-1}}(2u + v) > 1$, $\frac{1}{3^{m-1}}(2u - v) > 1$.

Hence, for every value of $m = 0, 1, 2, \ldots, n - 1$, we obtain

$$\frac{1}{3^n}(u) > 1, \frac{1}{3^n}(v) > 1, \frac{1}{3^n}(2u + v) > 1, \frac{1}{3^n}(2u - v) > 1$$

and $\Delta_1 m_t(3^{-n}u, 3^{-n}v) = 0$ for $m = 0, 1, 2, \ldots, n - 1$. Applying (2.9) and the definition of m_t, we obtain

$$|\Delta_1 m_t(u, v)| \leq \sum_{m=n}^{\infty} \frac{c}{1594323^m} + \sum_{m=n}^{\infty} \frac{c}{1594323^m} + \frac{1594324}{1594323} \sum_{m=n}^{\infty} \frac{c}{1594323^m}$$

$$\leq \frac{4782970\,c}{1594323} \frac{1}{1594323^m} \left(1 - \frac{1}{1594323}\right)^{-1}$$

$$\leq \frac{4782970\,c}{1594322} \frac{1}{1594323^{m+1}}$$

$$\leq \frac{2391485\,c}{797161} \left(|u|^{-13} + |v|^{-13}\right)$$

for all $u, v \in \mathbb{R}^*$. This means that the inequality (2.11) holds. We claim that the multiplicative inverse tredecic functional equation (2.1) is not stable for $\beta = -13$ in Corollary 2.3.4. Assume that there exists a multiplicative inverse tredecic mapping $m_t : \mathbb{R}^* \longrightarrow \mathbb{R}$ satisfying (2.12). So, we have

$$|m_t(u)| \leq (\delta + 1)|u|^{-13}. \tag{2.14}$$

Moreover, it is possible to choose a positive integer m with the condition $mc > \delta + 1$. If $u \in \left(1, 3^{m-1}\right)$, then $3^{-n}u \in (1, \infty)$ for all $m = 0, 1, 2, \ldots, n - 1$ and thus

$$|m_t(u)| = \sum_{m=0}^{\infty} \frac{\chi(3^{-m}u)}{1594323^m} \geq \sum_{m=0}^{n-1} \frac{\frac{1594323^m c}{u^{13}}}{1594323^m} = \frac{mc}{u^{13}} > (\delta + 1)u^{-13}$$

which contradicts (2.14). Therefore, the multiplicative inverse tredecic functional equation (2.1) is not stable for $\beta = -13$ in Corollary 2.3.4.

Similar to Example 2.4.1, the following example acts as a counter-example that the Eq. (2.2) is not stable for $\beta = -14$ in Corollary 2.3.9.

Example 2.4.2 Define the function $\xi : \mathbb{R}^* \longrightarrow \mathbb{R}$ via

$$\xi(u) = \begin{cases} \frac{\lambda}{u^{14}} & \text{for } u \in (1, \infty) \\ c, & \text{otherwise} \end{cases}. \tag{2.15}$$

Let $m_q : \mathbb{R}^* \longrightarrow \mathbb{R}$ be defined by

$$m_q(u) = \sum_{m=0}^{\infty} 4782969^{-m} \xi(3^{-m} s)$$

for all $u \in \mathbb{R}$. Suppose the function m_q satisfies the functional inequality

$$\left| \Delta_2 m_q(u, v) \right| \leq \frac{3587227 \, \lambda}{1195742} \left(|u|^{-14} + |v|^{-14} \right)$$

for all $u, v \in \mathbb{R}^*$. Then, there do not exist a multiplicative inverse tredecic mapping $m_Q : \mathbb{R}^* \longrightarrow \mathbb{R}$ and a constant $\eta > 0$ such that

$$\left| m_q(u) - m_Q(u) \right| \leq \eta \, |u|^{-14}$$

for all $u \in \mathbb{R}^*$.

Chapter 3
Estimation of Inexact Multiplicative Inverse Type Quindecic and Sexdecic Functional Equations in Felbin's Type Fuzzy Normed Spaces

Abstract This chapter is devoted to demonstrate the validation of various stabilities of multiplicative inverse quindecic and multiplicative inverse sexdecic functional equations via fixed point technique in the framework of Felbin's type fuzzy normed spaces. Proper illustrations are presented to disprove the stability results for singular cases.

3.1 Introduction

The fuzzy real number is considered as a decreasing mapping from the set of real numbers to [0, 1] by Gantner et al. [37]. Rodabaugh [124] studied the fuzzy arithmetic operations on L-fuzzy real line. The idea of fuzzy normed linear spaces was first introduced by Felbin [34] and its linear topological structures and some fundamental properties were studied by Xiao and Zhu [146]. There are many applications of fuzzy numbers intuitionistic fuzzy numbers in system design.

Shukla et al. [137] proved some fixed point theorems in 1-M-complete fuzzy metric like spaces and extended their theorems to a more general framework. These extended results are applied to generalize some well-known results. Stability analysis is an important aspect to investigate the qualitative analysis of solutions of dynamical systems with uncertainty. Fuzzy game theory is applied in many decision-making problems. Xia [145] investigated the matrix game with interval-valued intuitionistic fuzzy numbers based on Archimedean t-conorm and t-norm. The method proposed by Xia [145] provides more choices for players. Tao and Zhu [141] introduced two concepts of stability and attractivity in optimistic value for dynamical systems with uncertainty.

Pedrycz [103] has pointed out applicational aspects of fuzzy relations equations in system analysis. Pedrycz [104] also proposed a number of structural enhancements of the relational architectures using fuzzy relational equations. Perfilieva and Gottwald [105] demonstrated the connection between the problems of interpolation and approximation of fuzzy functions with solvability of systems of fuzzy relation equations.

B. V. Senthil Kumar and H. Dutta, *Multiplicative Inverse Functional Equations*,
Studies in Systems, Decision and Control 289, https://doi.org/10.1007/978-3-030-45355-8_3

In this chapter, a multiplicative inverse quindecic functional equation

$$M_Q(2u + v) + M_Q(2u - v)$$

$$= \frac{4M_Q(u)M_Q(v)}{\left(4M_Q(v)^{2/15} - M_Q(u)^{2/15}\right)^{15}} \left[\frac{1}{2} \sum_{k=0}^{7} \binom{15}{2k} [M_Q(u)]^{2k/15} [M_Q(v)]^{(15-2k)/15}\right]$$

$$(3.1)$$

and a multiplicative inverse sexdecic functional equation.

$$M_S(2u + v) + M_S(2u - v)$$

$$= \frac{4M_S(u)M_S(v)}{\left(4M_S(v)^{1/8} - M_S(u)^{1/8}\right)^{16}} \left[\sum_{k=0}^{8} \binom{16}{2k} [M_S(u)]^{k/8} [M_S(v)]^{(16-2k)/16}\right]$$

$$(3.2)$$

are introduced. It is not hard to verify that the multiplicative inverse quindecic function $M_Q(u) = \frac{1}{u^{15}}$ and the multiplicative inverse sexdecic function $M_S(u) = \frac{1}{u^{16}}$ are solutions of the Eqs. (3.1) and (3.2), respectively. We solve various Ulam stability problems of the Eqs. (3.1) and (3.2) in Felbin's type fuzzy normed spaces via fixed point technique. Proper counter-examples are presented to invalidate the stability results for control cases.

3.2 Preliminaries

In this section, we recall some preliminaries related to the theory of fuzzy reals numbers and Felbin's type fuzzy normed linear spaces. The fundamental results in fixed point theory are also evoked to ascertain the stability results.

A fuzzy number is a fuzzy set on the real axis, i.e., a mapping $x : \mathbb{R} \to [0, 1]$ associating with each real number t and its grade of membership $\eta(t)$. A fuzzy number x is convex if $x(t) \geq \min(x(s), x(r))$ where $s \leq t \leq r$.

Let η be a fuzzy subset on $\mathbb{R}$, i.e., a mapping $\eta : \mathbb{R} \to [0, 1]$ associating with each real number t its grade of membership $\eta(t)$.

Definition 3.2.1 ([34]) A fuzzy subset η on $\mathbb{R}$ is called a fuzzy real number, whose β−level set is denoted by $[\eta]_\beta$, i.e., $[\eta]_\beta = \{t : \eta(t) \geq \beta\}$, if it satisfies two axioms:

(M1) There exists $t_0 \in \mathbb{R}$ such that $\eta(t_0) = 1$.
(M2) For each $\beta \in (0, 1]$, $[\eta]_\beta = \left[\eta_\beta^-, \eta_\beta^+\right]$ where $-\infty < \eta_\beta^- \leq \eta_\beta^+ < +\infty$.

The set of all fuzzy real numbers is denoted by $f(\mathbb{R})$. If $\eta \in F(\mathbb{R})$, and $\eta(t) = 0$ whenever $t < 0$, then η is called a non-negative fuzzy real number, and $F^*(\mathbb{R})$ denotes the set of all non-negative fuzzy real numbers.

The number $\bar{0}$ stands for the fuzzy real number as:

$$\bar{0}(t) = \begin{cases} 1, t = 0, \\ 0, t \neq 0. \end{cases}$$

Clearly, $\bar{0} \in F^*(\mathbb{R})$. Also the set of all real numbers can be embeded in $f(\mathbb{R})$ because if $r \in (-\infty, \infty)$, then $\bar{r} \in F(\mathbb{R})$ satisfies $\bar{r}(t) = \bar{0}(t - r)$.

Definition 3.2.2 ([146]) Let E be a real linear space; L and R (respectively, left norm and right norm) be symmetric and non-decreasing mappings from $[0, 1] \times [0, 1]$ into $[0, 1]$ satisfying $L(0, 0) = 0$, $R(1, 1) = 1$. Then $\|\cdot\|$ is called a fuzzy norm, and $(E, \|\cdot\|, L, R)$ is a fuzzy normed linear space (abbreviated to FNLS) if the mapping $\|\cdot\|$ from E into $F^*(\mathbb{R})$ satisfies the following axioms, where $[\|u\|]_\beta = [\|u\|_\beta^-, \|u\|_\beta^+]$ for $u \in X$ and $\beta \in (0, 1]$:

(A1) $\|u\| = \bar{0}$ if and only if $u = 0$,
(A2) $\|\lambda u\| = |\lambda| \odot \|u\|$ for all $u \in E$ and $\lambda \in (-\infty, \infty)$,
(A3) For all $u, v \in E$:
(A3L) if $p \leq \|u\|_1^-, q \leq \|v\|_1^-$ and $p + q \leq \|p + q\|_1^-$, then
$\|u + v\|(p + q) \geq L(\|u\|(p), \|v\|(q))$,
(A3R) if $p \geq \|u\|_1^-, q \geq \|v\|_1^-$ and $p + q \geq \|u + v\|_1^-$, then
$\|u + v\|(p + q) \leq R(\|u\|(p), \|v\|(q))$.

Lemma 3.2.3 ([147]) *Let* $(E, \|\cdot\|, L, R)$ *be an FNLS, and suppose that*
$(R - 1)$ $R(u, v) \leq max(u, v)$,
$(R - 2)$ $\forall \beta \in (0, 1], \exists \mu \in (0, \beta]$ *such that* $R(\mu, v) \leq \beta$ *for all* $v \in (0, \beta)$,
$(R - 3)$ $\lim\limits_{a \to 0^+} R(a, a) = 0$.
Then $(R - 1) \Rightarrow (R - 2) \Rightarrow (R - 3)$ *but not conversely.*

Definition 3.2.4 ([129]) Let $(X, \|\cdot\|, L, R)$ be an FNLS. A sequence $\{x_n\}_{n=1}^\infty \subseteq X$ converges to $x \in X$, denoted by $\lim\limits_{n \to \infty} x_n = x$, if $\lim\limits_{n \to \infty} \|x_n - x\|_\alpha^+ = 0$ for every $\alpha \in (0, 1]$, and is called a Cauchy sequence if $\lim\limits_{m,n \to \infty} \|x_m - x_n\|_\alpha^+ = 0$ for every $\alpha \in (0, 1]$. A subset $A \subseteq X$ is said to be complete if every Cauchy sequence in A, converges in A. The fuzzy normed space $(X, \|\cdot\|, L, R)$ is said to be a fuzzy Banach space if it is complete.

Theorem 3.2.5 (Banach's contraction principle) *Let* (X, d) *be a complete metric and let* $\sigma : X \longrightarrow X$ *be strictly contractive, that is,*

$$d(\sigma u, \sigma v) \leq Ld(u, v), \quad \textit{for all } u, v \in X$$

for some Lipschitz constant $L < 1$*. Then*

1. the mapping σ *has a unique fixed point* $u^* = \sigma u^*$*;*

2. *the fixed point x^* is globally attractive, that is,*

$$\lim_{m \to \infty} \sigma^m u = u^*$$

for any starting point $u \in X$;
3. *one has the following estimation inequalities:*

$$d\left(\sigma^m u, u^*\right) \leq L^m d(u, u^*),$$

$$d\left(\sigma^m u, u^*\right) \leq \frac{1}{1-L} d\left(\sigma^m u, \sigma^{m+1} u\right),$$

$$d(u, u^*) \leq \frac{1}{1-L} d(u, \sigma u)$$

for all non-negative integers m and all $u \in X$.

Definition 3.2.6 Let E be a set. A function $d : E \times E \longrightarrow [0, \infty]$ is called a generalized metric on E if d satisfies

(1) $d(u, v) = 0$ if and only if $u = v$;
(2) $d(x, y) = d(y, x)$ for all $u, v \in E$;
(3) $d(u, w) \leq d(u, v) + d(v, w)$ for all $u, v, w \in E$.

Theorem 3.2.7 *Let (E, d) be a complete generalized metric space and let $\sigma : E \longrightarrow E$ be a strictly contractive mapping with Lipschitz constant $L < 1$. Then for each given element $u \in E$, either*

$$d\left(\sigma^m u, \sigma^{m+1} u\right) = \infty$$

for all non-negative integers n or there exists a positive integer n_0 such that

(1) $d\left(\sigma^m u, \sigma^{m+1} u\right) < \infty$,   *for all $m \geq m_0$;*
(2) *the sequence $\{\sigma^m u\}$ converges to a fixed point v^* of σ;*
(3) *v^* is the unique fixed point of σ in the set $\mathcal{V} = \{v \in E | d\left(\sigma^{m_0} u, v\right) < \infty\}$;*
(4) *$d(v, v^*) \leq \frac{1}{1-L} d(v, \sigma v)$ for all $v \in \mathcal{V}$.*

Further details about Felbin's type normed spaces and fixed point theory are available in [10, 17, 25, 146, 147].

3.3 Fuzzy Stabilities of Eqs. (3.1) and (3.2) via Fixed Point Method

In this section, we achieve various stabilities concerning Eqs. (3.1) and (3.2) through fixed point method in the setting of Felbin's type fuzzy normed spaces. Throughout this section, let us assume that E be a linear space and $(F, \|\cdot\|, L, R)$ be a fuzzy

Banach space satisfying $(R - 1)$. For the sake of simplicity, let us define the difference operators $\Delta_1, \Delta_2 : E \times E \longrightarrow F$, respectively, via

$$
\begin{aligned}
&\Delta_1 M_Q(u, v) \\
&\quad = M_Q(2u + v) + M_Q(2u - v) \\
&\qquad - \frac{4 M_Q(u) M_Q(v)}{\left(4 M_Q(v)^{2/15} - M_Q(u)^{2/15}\right)^{15}} \left[\frac{1}{2} \sum_{k=0}^{7} \binom{15}{2k} [M_Q(u)]^{2k/15} [M_Q(v)]^{(15-2k)/15} \right]
\end{aligned}
$$

and

$$
\begin{aligned}
&\Delta_2 M_S(u, v) \\
&\quad = M_S(2u + v) + M_S(2u - v) \\
&\quad = \frac{4 M_S(u) M_S(v)}{\left(4 M_S(v)^{1/8} - M_S(u)^{1/8}\right)^{16}} \left[\sum_{k=0}^{8} \binom{16}{2k} [M_S(u)]^{k/8} [M_S(v)]^{(16-2k)/16} \right]
\end{aligned}
$$

for all $u, v \in E$.

Theorem 3.3.1 *Let $\psi : E \times E \longrightarrow F^*(\mathbb{R})$ be a mapping satisfying*

$$
\lim_{p \to \infty} 3^{-15p} \psi \left(3^{-p} u, 3^{-p} v\right)_\beta^+ = 0 \tag{3.3}
$$

and let $M_Q : E \longrightarrow F$ be a mapping such that

$$
\left\| \Delta_1 M_Q (u, v) \right\| \preceq \psi(u, v) \tag{3.4}
$$

for all $u, v \in E$, and all $\beta \in (0, 1]$. Assume that there exists an $L < 1$ such that

$$
\psi \left(\frac{u}{3}, \frac{v}{3}\right) \preceq \frac{L}{3^{15}} \odot \psi(u, v)
$$

for all $u, v \in E$. Then there exists a unique multiplicative inverse quindecic mapping $\mathcal{M}_q : E \longrightarrow F$ satisfying (3.1) and

$$
\left\| M_Q(u) - \mathcal{M}_q(u) \right\|_\beta^+ \preceq \frac{1}{1 - L} \psi \left(\frac{u}{3}, \frac{u}{3}\right)_\beta^+
$$

for all $u \in E$ and all $\beta \in (0, 1]$.

Proof Let us define a set

$$
H = \{s : E \longrightarrow F\}
$$

and a generalized metric on H be introduced as follows:

$$
d(s, t) = \inf\{\lambda \in \mathbb{R}_+ : \|s(u) - t(u)\|_\beta^+ \leq \lambda \psi(u)_\beta^+, \forall u \in E, \forall \beta \in (0, 1)\}.
$$

It is easy to prove that (H, d) is complete. Now, consider an operator $\theta : H \longrightarrow H$ defined via

$$(\theta s)(u) = \frac{1}{3^{15}} s\left(\frac{u}{3}\right)$$

for all $u \in E$.

First, we prove that θ is strictly contractive on H. For any $s, t \in E$, and let $\lambda \in [0, \infty]$ be an arbitrary constant with $d(s, t) \leq \lambda$, that is,

$$d(s, t) < \lambda \Longrightarrow \|s(u) - t(u)\|_\beta^+ \leq \lambda \psi(u, u)_\beta^+, \forall u \in E, \forall \beta \in (0, 1]$$

$$\Longrightarrow \left\|\frac{1}{3^{15}} s\left(\frac{u}{3}\right) - \frac{1}{3^{15}} t\left(\frac{u}{3}\right)\right\|_\beta^+ \leq \frac{\lambda}{3^{15}} \psi\left(\frac{u}{3}, \frac{u}{3}\right)_\beta^+$$

$$\Longrightarrow \left\|\frac{1}{3^{15}} s\left(\frac{u}{3}\right) - \frac{1}{3^{15}} t\left(\frac{u}{3}\right)\right\|_\beta^+ \leq \lambda L \psi(u, u)_\beta^+$$

$$\Longrightarrow d(\theta s, \theta t) \leq \lambda L.$$

Hence we observe that $d(\theta s, \theta t) \leq L d(s, t)$, $\forall s, t \in H$. This implies that θ is strictly contractive self-mapping of H, with the Lipschitz constant L.

Now, replacing (uv) by $\left(\frac{u}{3}, \frac{u}{3}\right)$ in (3.4), then (3.4) produces

$$\left\|M_Q(u) - \frac{1}{3^{15}} M_Q\left(\frac{u}{3}\right)\right\| \preceq \psi\left(\frac{u}{3}, \frac{u}{3}\right)$$

for all $u \in E$. Hence, we conclude that $d(M_Q, \theta M_Q) \leq 1$.

The fundamental theorem of fixed point theory implies that there exists a mapping $\mathcal{M}_q : E \longrightarrow F$ such that

(1) $\mathcal{M}_q$ is a fixed point of θ, that is,

$$\mathcal{M}_q\left(\frac{u}{3}\right) = 3^{15} \mathcal{M}_q(u)$$

for all $u \in E$. Also, the mapping $\mathcal{M}_q$ is a unique fixed point of θ in the set

$$\Lambda = \{s \in E : d(s, t) < \infty\}.$$

From this, we see that $\mathcal{M}_q$ is a unique mapping such that there exists $\lambda \in (0, \infty)$ satisfying

$$\left\|\mathcal{M}_q(u) - M_Q(u)\right\|_\beta^+ \leq \lambda \psi(u, u)$$

for all $u \in E$.

(2) $d\left(\theta^m M_Q, \mathcal{M}_q\right) \longrightarrow 0$ as $m \longrightarrow \infty$, which implies the inequality

$$\lim_{m \to \infty} 3^{-15m} M_Q\left(3^{-m} u\right) = \mathcal{M}_q(u) \tag{3.5}$$

for all $u \in E$.

(3) $d(M_Q, \mathcal{M}_q) \leq \frac{1}{1-L} d(M_Q, \theta\mathcal{M}_q)$, which implies the inequality

$$d(M_Q, \mathcal{M}_q) \leq \frac{1}{1-L}.$$

It follows from (3.3) and (3.5), that

$$\begin{aligned}
\left\| \Delta_1 \mathcal{M}_q(u, v) \right\|_\beta^+ &= \lim_{m \to \infty} 3^{-15m} \left\| \Delta_1 M_Q(u, v) \right\|_\beta^+ \\
&\leq \lim_{m \to \infty} 3^{-15m} \psi \left(3^{-m}u, 3^{-m}v \right)_\beta^+ = 0
\end{aligned}$$

for all $u, v \in E$. Hence the mapping $\mathcal{M}_q$ is multiplicative inverse quindecic which completes the proof.

Corollary 3.3.2 *Suppose ϵ is a non-negative fuzzy real number. If $M_Q : E \longrightarrow F$ is a mapping satisfying*

$$\left\| \Delta_1 M_Q(u, v) \right\| \preceq \epsilon$$

for all $u, v \in E$, then there exists a unique multiplicative inverse quindecic mapping $\mathcal{M}_q : E \longrightarrow F$ satisfying (3.1) and

$$\left\| M_Q(u) - Q_q(u) \right\|_\beta^+ \preceq \frac{3^{15}}{3^{15} - 1} \epsilon_\beta^+$$

for all $u \in E$.

Proof Define $\psi(u, v) = \epsilon$ in Theorem 3.3.1 and assume $L = \frac{1}{3^{15}}$, we get the requisite result. □

Corollary 3.3.3 *Let ϵ be a non-negative fuzzy real number. Let $a(> -15)$ be a real number. If $M_Q : E \longrightarrow F$ is a mapping satisfying*

$$\left\| \Delta_1 M_Q(u, v) \right\|_F \preceq \epsilon \otimes \left(\|u\|_E^a \oplus \|v\|_E^a \right)$$

for all $u, v \in E$, then there exists a unique multiplicative inverse quindecic mapping $Q_q : E \longrightarrow F$ satisfying (3.1) and

$$\left\| M_Q(u) - Q_q(u) \right\|_\beta^+ \preceq \frac{2 \cdot 3^{15}}{3^{a+15} - 1} \epsilon_\beta^+ \otimes \left(|u|_E^a \right)$$

for all $u \in E$, and all $\beta \in (0, 1]$.

Proof Choosing $\psi(u, v) = \epsilon \otimes \left(\|u\|_E^a \oplus \|v\|_E^a \right)$ and $L = \frac{1}{3^{a+15}}$ in Theorem 3.3.1, we obtain the desired result. □

Corollary 3.3.4 *Let ϵ be a non-negative fuzzy real number. Let k, ℓ be real numbers such that $\rho = k + \ell > -15$. If a mapping $M_Q : E \longrightarrow F$ is a mapping satisfying*

$$\left\| \Delta_1 M_Q(u, v) \right\| \preceq \epsilon \otimes \left(\|u\|_E^k \otimes \|v\|_E^\ell \right)$$

for all $u, v \in E$, then there exists unique multiplicative inverse quindecic mapping $\mathcal{M}_q : E \longrightarrow F$ satisfying (3.1) and

$$\left\| M_Q(u) - \mathcal{Q}_q(u) \right\|_\beta^+ \leq \frac{3^{15}}{3^{k+\ell} \left(3^{k+\ell+15} - 1 \right)} \left(\epsilon \otimes \|u\|_E^{k+\ell} \right)$$

for all $u \in E$, and all $\beta \in (0, 1]$.

Proof Taking $\psi(u, v) = \epsilon \otimes \left(\|u\|_E^k \otimes \|v\|_E^\ell \right)$ and $L = \frac{1}{3^{k+\ell+15}}$ in Theorem 3.3.1, we arrive at the required result.   □

By the application of fixed point method, the fundamental stabilities of Eq. (3.2) are investigated in non-Archimedean fields. We omit the proof of the stability results of Eq. (3.2) as they are obtained with similar arguments as in the results associated with Eq. (3.1). Hence only statement of theorems are presented without their proofs.

Theorem 3.3.5 *Let $\psi : E \times E \longrightarrow F^*(\mathbb{R})$ be a mapping satisfying*

$$\lim_{p \to \infty} 3^{-16p} \psi \left(3^{-p} u, 3^{-p} v \right)_\beta^+ = 0$$

and let $M_S : E \longrightarrow F$ be a mapping such that

$$\left\| \Delta_2 M_S(u, v) \right\| \preceq \psi(u, v)$$

for all $u, v \in E$, and all $\beta \in (0, 1]$. Assume that there exists an $L < 1$ such that

$$\psi \left(\frac{u}{3}, \frac{v}{3} \right) \preceq \frac{L}{3^{16}} \odot \psi(u, v)$$

for all $u, v \in E$. Then there exists a unique multiplicative inverse sexdecic mapping $\mathcal{M}_s : E \longrightarrow F$ satisfying (3.2) and

$$\left\| M_S(u) - \mathcal{M}_s(u) \right\|_\beta^+ \leq \frac{1}{1 - L} \psi \left(\frac{u}{3}, \frac{u}{3} \right)_\beta^+$$

for all $u \in E$ and all $\beta \in (0, 1]$.

Corollary 3.3.6 *Suppose ϵ is a non-negative fuzzy real number. If $M_S : E \longrightarrow F$ is a mapping satisfying*

$$\left\| \Delta_2 M_S(u, v) \right\| \preceq \epsilon$$

for all $u, v \in E$, then there exists a unique multiplicative inverse sexdecic mapping $\mathcal{M}_s : E \longrightarrow F$ satisfying (3.2) and

$$\|M_S(u) - \mathcal{M}_s(u)\|_\beta^+ \preceq \frac{3^{16}}{3^{16}-1}\epsilon_\beta^+$$

for all $u \in E$.

Corollary 3.3.7 *Let ϵ be a non-negative fuzzy real number. Let $a(> -16)$ be a real number. If $M_S : E \longrightarrow F$ is a mapping satisfying*

$$\|\Delta_2 M_S(u, v)\|_F \preceq \epsilon \otimes \left(\|u\|_E^a \oplus \|v\|_E^a\right)$$

for all $u, v \in E$, then there exists a unique multiplicative inverse sexdecic mapping $\mathcal{Q}_q : E \longrightarrow F$ satisfying (3.2) and

$$\|M_S(u) - \mathcal{M}_s(u)\|_\beta^+ \preceq \frac{2 \cdot 3^{16}}{3^{a+16}-1}\epsilon_\beta^+ \otimes \left(|u|_E^a\right)$$

for all $u \in E$, and all $\beta \in (0, 1]$.

Corollary 3.3.8 *Let ϵ be a non-negative fuzzy real number. Let k, ℓ be real numbers such that $\rho = k + \ell > -16$. If a mapping $M_S : E \longrightarrow F$ is a mapping satisfying*

$$\|\Delta_2 M_S(u, v)\|_F \preceq \epsilon \otimes \left(\|u\|_E^k \otimes \|v\|_E^\ell\right)$$

for all $u, v \in E$, then there exists unique multiplicative inverse sexdecic mapping $\mathcal{M}_s : E \longrightarrow F$ satisfying (3.2) and

$$\|M_S(u) - \mathcal{M}_s(u)\|_\beta^+ \preceq \frac{3^{16}}{3^{k+\ell}\left(3^{k+\ell+16}-1\right)}\left(\epsilon \otimes \|u\|_E^k\right)$$

for all $u \in E$, and all $\beta \in (0, 1]$.

3.4 Counter-Examples

In this section, we show that the Eqs. (3.1) and (3.2) are not valid for $a = -15$ in Corollary 3.3.3 and $a = -16$ in Corollary 3.3.7, respectively, in the setting of non-zero real numbers.

Example 3.4.1 Let us define the function

$$\chi(u) = \begin{cases} \frac{c}{u^{15}}, & \text{for } u \in (1, \infty) \\ c, & \text{elsewhere} \end{cases} \tag{3.6}$$

where $\chi : \mathbb{R}^* \longrightarrow \mathbb{R}$. Let $M_Q : \mathbb{R}^* \longrightarrow \mathbb{R}$ be a function defined as

$$M_Q(u) = \sum_{m=0}^{\infty} 14348907^{-m} \chi(3^{-m}u) \tag{3.7}$$

for all $u \in \mathbb{R}$. Suppose the mapping $M_Q : \mathbb{R}^* \longrightarrow \mathbb{R}$ described in (3.7) satisfies the inequality

$$\left| \Delta_1 M_Q(u, v) \right| \le \frac{21523361\, c}{7174453} \left(|u|^{-15} + |v|^{-15} \right) \tag{3.8}$$

for all $u, v \in \mathbb{R}^*$. We prove that there do not exist a multiplicative inverse quindecic mapping $\mathcal{M}_q : \mathbb{R}^* \longrightarrow \mathbb{R}$ and a constant $\delta > 0$ such that

$$\left| M_Q(u) - \mathcal{M}_q(u) \right| \le \delta\, |u|^{-15} \tag{3.9}$$

for all $u \in \mathbb{R}^*$. Firstly, let us prove that M_Q satisfies (3.8). Using (3.6), we have

$$\left| M_Q(u) \right| = \left| \sum_{m=0}^{\infty} 14348907^{-m} \chi(3^{-m}u) \right| \le \sum_{m=0}^{\infty} \frac{c}{14348907^m} = \frac{14348907}{14348906} c.$$

We observe that M_Q is bounded by $\frac{14348907\, c}{14348906}$ on $\mathbb{R}$. If $|u|^{-15} + |v|^{-15} \ge 1$, then the left hand side of (3.8) is less than $\frac{21523361\, c}{7174453}$. Now, suppose that $0 < |u|^{-15} + |v|^{-15} < 1$. Hence, there exists a positive integer m such that

$$\frac{1}{14348907^{m+1}} \le |u|^{-15} + |v|^{-15} < \frac{1}{14348907^m}. \tag{3.10}$$

Hence, the inequality (3.10) produces $14348907^m \left(|u|^{-15} + |v|^{-15} \right) < 1$, or equivalently; $14348907^m u^{-15} < 1$, $14348907^m v^{-15} < 1$. So, $\frac{u^{15}}{14348907^m} > 1$, $\frac{v^{15}}{14348907^m} > 1$. Hence, the last inequalities imply $\frac{u^{15}}{14348907^{m-1}} > 14348907 > 1$, $\frac{v^{15}}{14348907^{m-1}} > 14348907 > 1$ and as a result, we find $\frac{1}{3^{m-1}}(u) > 1$, $\frac{1}{3^{m-1}}(v) > 1$, $\frac{1}{3^{m-1}}(2u + v) > 1$, $\frac{1}{3^{m-1}}(2u - v) > 1$. Hence, for every value of $m = 0, 1, 2, \ldots, n - 1$, we obtain

$$\frac{1}{3^n}(u) > 1, \quad \frac{1}{3^n}(v) > 1, \quad \frac{1}{3^n}(2u + v) > 1, \quad \frac{1}{3^n}(2u - v) > 1$$

and $\Delta_1 M_Q(3^{-n}u, 3^{-n}v) = 0$ for $m = 0, 1, 2, \ldots, n - 1$. Applying (3.6) and the definition of M_Q, we obtain

$$\left|\Delta_1 M_Q(u, v)\right| \leq \sum_{m=n}^{\infty} \frac{c}{14348907^m} + \sum_{m=n}^{\infty} \frac{c}{14348907^m} + \frac{14348908}{14348907} \sum_{m=n}^{\infty} \frac{c}{14348907^m}$$

$$\leq \frac{43046722\, c}{14348907} \frac{1}{14348907^m} \left(1 - \frac{1}{14348907}\right)^{-1}$$

$$\leq \frac{43046722\, c}{14348906} \frac{1}{14348907^{m+1}}$$

$$\leq \frac{21523361\, c}{7174453} \left(|u|^{-15} + |v|^{-15}\right)$$

for all $u, v \in \mathbb{R}^*$. This means that the inequality (3.8) holds. We claim that the multiplicative inverse quindecic functional equation (3.1) is not stable for $a = -15$ in Corollary 3.3.3. Assume that there exists a multiplicative inverse quindecic mapping $M_Q : \mathbb{R}^* \longrightarrow \mathbb{R}$ satisfying (3.9). So, we have

$$|M_Q(u)| \leq (\delta + 1)|u|^{-15}. \tag{3.11}$$

Moreover, it is possible to choose a positive integer m with the condition $mc > \delta + 1$. If $u \in \left(1, 3^{m-1}\right)$, then $3^{-n}u \in (1, \infty)$ for all $m = 0, 1, 2, \ldots, n - 1$ and thus

$$|M_Q(u)| = \sum_{m=0}^{\infty} \frac{\chi(3^{-m}u)}{14348907^m} \geq \sum_{m=0}^{n-1} \frac{\frac{14348907^m c}{u^{15}}}{14348907^m} = \frac{mc}{u^{15}} > (\delta + 1)u^{-15}$$

which contradicts (3.11). Therefore, the multiplicative inverse quindecic functional equation (3.1) is not stable for $a = -15$ in Corollary 3.3.3.

Similar to Example 3.4.1, the following example acts as a counter-example that the Eq. (3.2) is not stable for $a = -16$ in Corollary 3.3.7.

Example 3.4.2 Define the function $\xi : \mathbb{R}^* \longrightarrow \mathbb{R}$ via

$$\xi(u) = \begin{cases} \frac{\lambda}{u^{16}} & \text{for } u \in (1, \infty) \\ c, & \text{otherwise} \end{cases}. \tag{3.12}$$

Let $M_S : \mathbb{R}^* \longrightarrow \mathbb{R}$ be defined by

$$M_S(u) = \sum_{m=0}^{\infty} 43046721^{-m} \xi(3^{-m}s)$$

for all $u \in \mathbb{R}$. Suppose the function M_S satisfies the inequality

$$|\Delta_2 M_S(u, v)| \leq \frac{32285041\, \lambda}{10761680} \left(|u|^{-16} + |v|^{-16}\right)$$

for all $u, v \in \mathbb{R}^*$. Then, there do not exist a multiplicative inverse sexdecic mapping $\mathcal{Q}_q : \mathbb{R}^* \longrightarrow \mathbb{R}$ and a constant $\eta > 0$ such that

$$\left| M_Q(u) - \mathcal{Q}_q(u) \right| \leq \eta \, |u|^{-16}$$

for all $u \in \mathbb{R}^*$.

Chapter 4
Classical Approximations of Multiplicative Inverse Type Septendecic and Octadecic Functional Equations in Quasi-β-normed Spaces

Abstract This chapter contains the classical investigation of various fundamental stability results of multiplicative inverse septendecic and octadecic functional equations in quasi-β-normed spaces using fixed point technique and also includes two proper examples to disprove stability results for control cases.

4.1 Introduction and Preliminaries

We elicit here, a few primitive ideas of quasi-β-normed spaces. Let β be a fixed real number with $0 < \beta \leq 1$ and let $\mathbb{K}$ denote either $\mathbb{R}$ or $\mathbb{C}$.

Definition 4.1.1 Let $\mathcal{X}$ be a linear space over $\mathbb{K}$. A quasi-β-norm $\|\cdot\|$ is a real-valued function on $\mathcal{X}$ satisfying the following conditions: (i) $\|u\| \geq 0, \forall u \in \mathcal{X}$ and $\|u\| = 0$ if and only if $u = 0$; (ii) $\|\mu u\| = |\mu|^{\beta} \cdot \|u\|, \forall \mu \in \mathbb{K}$ and $\forall u \in \mathcal{X}$; (iii) There is a constant $K > 1$ such that $\|u + v\| \leq K (\|u\| + \|v\|), \forall\, u, v \in \mathcal{X}$.

The pair $(\mathcal{X}, \|\cdot\|)$ is called quasi-β-normed space if $\|\cdot\|$ is a quasi-β-norm on $\mathcal{X}$. The smallest possible K is called the modulus of concavity of $\|\cdot\|$.

Definition 4.1.2 A quasi-β-Banach space is a complete quasi-β-normed space.

In this chapter, the fundamental stabilities of a multiplicative inverse septendecic functional equation

$$
m_s(2u + v) + m_s(2u - v)
$$
$$
= \frac{4m_s(u)m_s(v)}{\left(4m_s(v)^{2/17} - m_s(u)^{2/17}\right)^{17}} \left[\frac{1}{2} \sum_{k=0}^{8} \binom{17}{2k} [m_s(u)]^{2k/17} [m_s(v)]^{(17-2k)/17} \right]
$$

$$(4.1)$$

and a multiplicative inverse octadecic functional equation

© The Editor(s) (if applicable) and The Author(s), under exclusive license
to Springer Nature Switzerland AG 2020
B. V. Senthil Kumar and H. Dutta, *Multiplicative Inverse Functional Equations*,
Studies in Systems, Decision and Control 289, https://doi.org/10.1007/978-3-030-45355-8_4

$$m_o(2u + v) + m_o(2u - v)$$

$$= \frac{4m_o(u)m_o(v)}{\left(4m_o(v)^{1/9} - m_s(u)^{1/9}\right)^{18}} \left[\sum_{k=0}^{9} \binom{18}{2k} [m_o(u)]^{k/9} [m_s(v)]^{(18-2k)/18} \right] \quad (4.2)$$

are discussed in the framework of quasi-β-normed spaces. Suitable counter-examples are provided to illustrate the non-stability of the Eqs. (4.1) and (4.2) for singular cases.

Throughout this chapter, let $\mathcal{X}$ be a quasi-β-normed space and let $\mathcal{Y}$ be a quasi-β-Banach space with a quasi-β-norm $\|\cdot\|_{\mathcal{Y}}$. For a given mapping $m_s, m_o : \mathcal{X} \to \mathcal{Y}$, let us define the difference operators $D_1 m_s, D_2 m_o : \mathcal{X} \times \mathcal{X} \longrightarrow \mathcal{Y}$ by

$$D_1 m_s(u, v)$$
$$= m_s(2u + v) + m_s(2u - v)$$
$$- \frac{4m_s(u)m_s(v)}{\left(4m_s(v)^{2/17} - m_s(u)^{2/17}\right)^{17}} \left[\frac{1}{2} \sum_{k=0}^{8} \binom{17}{2k} [m_s(u)]^{2k/17} [m_s(v)]^{(17-2k)/17} \right]$$

and

$$D_2 m_o(u, v)$$
$$= m_o(2u + v) + m_o(2u - v)$$
$$- \frac{4m_o(u)m_o(v)}{\left(4m_o(v)^{1/9} - m_s(u)^{1/9}\right)^{18}} \left[\sum_{k=0}^{9} \binom{18}{2k} [m_o(u)]^{k/9} [m_s(v)]^{(18-2k)/18} \right]$$

for all $y_1, y_2, \ldots, y_n \in \mathcal{X}$.

4.2 Fundamental Stabilities of Eqs. (4.1) and (4.2)

In this section, we discuss about the fundamental stability of Eqs. (4.1) and (4.2) pertinent to Gavruta and then extend to the stability results relevant to Hyers, T. Rassias and J. Rassias in quasi-β-normed spaces.

Theorem 4.2.1 *Assume $\psi : \mathcal{X} \times \mathcal{X} \longrightarrow [0, \infty)$ is a mapping satisfying the following inequality:*

$$\sum_{j=0}^{\infty} \left(\frac{K}{3^{17\beta}} \right)^j \psi \left(\frac{u}{3^{j+1}}, \frac{v}{3^{j+1}} \right) < \infty \quad (4.3)$$

for all $u, v \in \mathcal{X}$. Suppose $m_s : \mathcal{X} \longrightarrow \mathcal{Y}$ is a mapping which satisfies the ensuing inequality

$$\|D_1 m_s(u, v)\|_{\mathcal{Y}} \le \psi(u, v) \tag{4.4}$$

for all $u, v \in \mathcal{X}$. Then, there exists a unique multiplicative inverse septendecic mapping $\mathcal{M}_S : \mathcal{X} \longrightarrow \mathcal{Y}$ satisfying (4.1) and

$$\|m_s(u) - \mathcal{M}_S(u)\|_{Y} \le K \sum_{j=0}^{\infty} \left(\frac{K}{3^{17\beta}}\right)^j \psi\left(\frac{u}{3^{j+1}}, \frac{u}{3^{j+1}}\right) \tag{4.5}$$

for all $u \in \mathcal{X}$. The mapping $\mathcal{M}_S$ is defined as

$$\mathcal{M}_S(u) = \lim_{m \to \infty} \frac{1}{3^{17m}} m_s\left(\frac{u}{3^m}\right) \tag{4.6}$$

for all $y \in \mathcal{X}$.

Proof Switching (u, v) to $\left(\frac{u}{3}, \frac{u}{3}\right)$ in (4.4) and simplifying further, we obtain

$$\left\| m_s(u) - \frac{1}{3^{17}} m_s\left(\frac{u}{3}\right) \right\|_{y} \le \psi\left(\frac{u}{3}, \frac{u}{3}\right) \tag{4.7}$$

for all $u \in \mathcal{X}$. Now, replacing u by $\frac{u}{3}$, dividing by $3^{17\beta}$ in (4.7) and summing the resulting inequality with (4.7), we find

$$\left\| m_s(u) - \frac{1}{3^{34}} m_s\left(\frac{u}{3^2}\right) \right\|_{y} \le K \sum_{j=1}^{1} \left(\frac{K}{n^{17\beta}}\right)^j \psi\left(\frac{u}{3^{j+1}}, \frac{u}{3^{j+1}}\right)$$

for all $u \in \mathcal{X}$. Proceeding further in the similar fashion and applying induction arguments on an integer $m > 0$, we arrive

$$\left\| m_s(u) - \frac{1}{3^{17m}} m_s\left(\frac{u}{3^m}\right) \right\|_{y} \le K \sum_{j=1}^{m-1} \left(\frac{K}{3^{17\beta}}\right)^j \psi\left(\frac{u}{3^{j+1}}, \frac{u}{3^{j+1}}\right) \tag{4.8}$$

for all $u \in \mathcal{X}$. From (4.7), we obtain

$$\left\| \frac{1}{3^{17j}} m_s\left(\frac{u}{3^j}\right) - \frac{1}{3^{17(j+1)}} m_s\left(\frac{u}{3^{j+1}}\right) \right\|_{y} \le \frac{1}{3^{17j\beta}} \psi\left(\frac{u}{3^{j+1}}, \frac{u}{3^{j+1}}\right) \tag{4.9}$$

for all $u \in \mathcal{X}$. For $m > l$,

$$\left\| \frac{1}{3^{17m}} m_s \left(\frac{u}{3^m} \right) - \frac{1}{3^{17l}} m_s \left(\frac{u}{3^l} \right) \right\|_{\mathcal{Y}} \leq \sum_{j=l}^{m-1} \left\| \frac{1}{3^{17(j+1)}} m_s \left(\frac{u}{3^{j+1}} \right) - \frac{1}{3^{17j}} m_s \left(\frac{u}{3^j} \right) \right\|_{\mathcal{Y}}$$

$$\leq \sum_{j=l}^{m-1} \frac{1}{3^{17j\beta}} \psi \left(\frac{u}{3^{j+1}}, \frac{u}{3^{j+1}} \right) \quad (4.10)$$

for all $u \in \mathcal{X}$. It is easy to find that the right-hand side of the above inequality (4.10) approaches to 0 as $m \to \infty$ which implies that the sequence $\left\{ \frac{1}{3^{17m}} m_s \left(\frac{u}{3^m} \right) \right\}$ is a Cauchy sequence in $\mathcal{Y}$. Hence, we can define

$$\mathcal{M}_S(u) = \lim_{m \to \infty} \frac{1}{3^{17m}} m_s \left(\frac{u}{3^m} \right)$$

for all $u \in \mathcal{X}$. In lieu of $K \geq 1$, switching (u, v) to $\left(\frac{u}{3^m}, \frac{v}{3^m} \right)$ and then dividing by $3^{17m\beta}$ in (4.4), we have

$$\frac{1}{3^{17m\beta}} \left\| D_1 m_s \left(\frac{u}{3^m}, \frac{v}{3^m} \right) \right\|_{\mathcal{Y}} \leq \frac{K^m}{3^{17m\beta}} \psi \left(\frac{u}{3^m}, \frac{v}{3^m} \right) \quad (4.11)$$

for all $u, v \in \mathcal{X}$. Allowing $m \to \infty$ in (4.11) and using the definition of $\mathcal{M}_S$, we observe that $\mathcal{M}_S$ satisfies (4.1) for all $u, v \in \mathcal{X}$. Hence $\mathcal{M}_S$ is a multiplicative inverse septendecic mapping. Also, the inequality (4.8) produces the inequality (4.5). Next is to show the uniqueness of $\mathcal{M}_S$. Assume that there exists $\mathcal{M}'_S : \mathcal{X} \longrightarrow \mathcal{Y}$ satisfying (4.1) and (4.5). It is easy to show that for all $u \in \mathcal{X}$, $\mathcal{M}'_S \left(\frac{u}{3^m} \right) = 3^{17m} \mathcal{M}'_S(u)$ and $\mathcal{M}_S \left(\frac{u}{3^m} \right) = 3^{17m} \mathcal{M}_S(u)$. Then

$$\left\| \mathcal{M}'_S(u) - \mathcal{M}_S(u) \right\|_{\mathcal{Y}}$$

$$= \left\| \frac{1}{3^{17m}} \mathcal{M}'_S \left(\frac{u}{3^m} \right) - \frac{1}{3^{17m}} \mathcal{M}_S \left(\frac{u}{3^m} \right) \right\|_{\mathcal{Y}}$$

$$= \frac{1}{3^{17m\beta}} \left\| \mathcal{M}'_S \left(\frac{u}{3^m} \right) - \mathcal{M}_S \left(\frac{u}{3^m} \right) \right\|_{\mathcal{Y}}$$

$$\leq \frac{K}{3^{17m\beta}} \left(\left\| \mathcal{M}'_S \left(\frac{u}{3^m} \right) - m_s \left(\frac{u}{3^m} \right) \right\|_{\mathcal{Y}} + \left\| m_s \left(\frac{u}{3^m} \right) - \mathcal{M}_S \left(\frac{u}{3^m} \right) \right\|_{\mathcal{Y}} \right)$$

$$\leq 2K^2 \sum_{j=0}^{\infty} \left(\frac{K}{3^{17\beta}} \right)^{m+j} \psi \left(\frac{u}{3^{m+j+1}}, \frac{u}{3^{m+j+1}} \right)$$

for all $u \in \mathcal{Y}$. By letting $m \to \infty$, we immediately have the uniqueness of $\mathcal{M}_S$. $\square$

The subsequent theorem is dual of Theorem 4.2.1. Hence, we present the main part of the proof as it is analogous to Theorem 4.2.1.

Theorem 4.2.2 *Let $\psi : \mathcal{X} \times \mathcal{X} \longrightarrow [0, \infty)$ be a mapping satisfying the subsequent inequality:*

$$\sum_{j=0}^{\infty} \left(3^{17\beta} K\right)^j \psi \left(3^j u, 3^j v\right) < \infty \tag{4.12}$$

for all $u, v \in \mathcal{X}$. Let $m_s : \mathcal{X} \longrightarrow \mathcal{Y}$ be a mapping satisfying (4.4) for all $u, v \in \mathcal{X}$. Then there exists a unique multiplicative inverse septendecic mapping $\mathcal{M}_S : \mathcal{X} \longrightarrow \mathcal{Y}$ satisfying (4.1) and

$$\|m_s(u) - \mathcal{M}_S(u)\|_{\mathcal{Y}} \le 3^{17\beta} K \sum_{j=0}^{\infty} \left(3^{17\beta} K\right)^j \psi \left(3^j u, 3^j v\right) \tag{4.13}$$

for all $u \in \mathcal{X}$. The mapping $\mathcal{M}_S$ is defined by

$$\mathcal{M}_S(u) = \lim_{m \to \infty} 3^{17m} m_s \left(3^m u\right) \tag{4.14}$$

for all $u \in \mathcal{X}$.

Proof Reinstating (u, v) into (u, u) in (4.4) and multiplying by $3^{17\beta}$, we obtain

$$\left\|3^{17} m_s(3u) - m_s(u)\right\|_{\mathcal{Y}} \le 3^{17\beta} \psi(u, u) \tag{4.15}$$

for all $u \in \mathcal{X}$. Now, replacing u as $3u$, multiplying by $3^{17\beta}$ in (4.15) and summing the resulting inequality with (4.15), we have

$$\left\|3^{34} m_s \left(3^2 u\right) - m_s(u)\right\|_{\mathcal{Y}} \le 3^{17\beta} K \sum_{j=0}^{1} \left(3^{17\beta} K\right)^j \psi \left(3^j u, 3^j v\right)$$

for all $u \in \mathcal{X}$. Using induction arguments on a positive integer m, we conclude that

$$\left\|3^{17m} m_s \left(3^m u\right) - m_s(u)\right\|_{\mathcal{Y}} \le 3^{17\beta} K \sum_{j=0}^{m-1} \left(3^{17\beta} K\right)^j \psi \left(3^j u, 3^j v\right)$$

for all $u \in \mathcal{X}$. The rest of the proof is analogous to Theorem 4.2.1. $\qquad \square$

The following corollaries are immediate consequences of Theorems 4.2.1 and 4.2.2 related to the results of Hyers, T. Rassias and J. Rassias.

Corollary 4.2.3 *Let $\epsilon \ge 0$ be fixed. If a mapping $m_s : \mathcal{X} \longrightarrow \mathcal{Y}$ satisfies the inequality*

$$\|D_1 m_s(u, v)\|_{\mathcal{Y}} \le \epsilon$$

for all $u, v \in \mathcal{X}$, then there exists a unique multiplicative inverse septendecic mapping $\mathcal{M}_S : \mathcal{X} \longrightarrow \mathcal{Y}$ satisfying (4.1) amd

$$\|m_s(u) - \mathcal{M}_S(u)\|_{\mathcal{Y}} \leq \frac{3^{17\beta} K \epsilon}{3^{17\beta} - K}$$

for all $u \in \mathcal{X}$.

Proof Considering $\psi(u, v) = \epsilon$ in Theorem 4.2.1, we arrive at the desired result. □

Corollary 4.2.4 *Let $\delta \geq 0$ be fixed and $\alpha \neq -17\beta$. If a mapping $m_s : \mathcal{X} \longrightarrow \mathcal{Y}$ satisfies the inequality*

$$\|D_1 m_s(u, v)\|_{\mathcal{Y}} \leq \delta \left(\|u\|_{\mathcal{X}}^{\alpha} + \|v\|_{\mathcal{X}}^{\alpha} \right)$$

for all $u, v \in \mathcal{X}$, then there exists a unique multiplicative inverse septendecic mapping $\mathcal{M}_S : \mathcal{X} \longrightarrow \mathcal{Y}$ satisfying (4.1) and

$$\|m_s(u) - \mathcal{M}_S(u)\|_{\mathcal{Y}} \leq \begin{cases} \frac{3^{17\beta+1} K \delta}{3^{17\beta+\alpha} - K} \|u\|_{\mathcal{X}}^{\alpha}, & \text{for } \alpha > -17\beta, \\ \frac{3^{17\beta+1} K \delta}{1 - 3^{17\beta+\alpha} K} \|u\|_{\mathcal{X}}^{\alpha}, & \text{for } \alpha < -17\beta \end{cases}$$

for all $u \in \mathcal{X}$.

Proof By choosing $\psi(u, v) = \delta \left(\|u\|_{\mathcal{X}}^{\alpha} + \|v\|_{\mathcal{X}}^{\alpha} \right)$ and $\alpha > -17\beta$ in Theorem 4.2.1 and $\alpha < -17\beta$ in Theorem 4.2.2, respectively, we obtain the required results. □

Corollary 4.2.5 *Let $\delta \geq 0$ be a fixed real number and α be such that $\alpha \neq -17\beta$. If a mapping $m_s : \mathcal{X} \longrightarrow \mathcal{Y}$ satisfies the inequality*

$$\|D_1 m_s(u, v)\|_{\mathcal{Y}} \leq \delta \|u\|_{\mathcal{X}}^{\alpha/2} \|v\|_{\mathcal{X}}^{\alpha/2}$$

for all $u, v \in \mathcal{X}$, then there exists a unique multiplicative inverse septendecic mapping $\mathcal{M}_S : \mathcal{X} \longrightarrow \mathcal{Y}$ satisfying (4.1) and

$$\|m_s(u) - \mathcal{M}_S(u)\|_{\mathcal{Y}} \leq \begin{cases} \frac{3^{17\beta} K \delta}{3^{17\beta+\alpha} - K} \|u\|_{\mathcal{X}}^{\alpha}, & \text{for } \alpha > -17\beta \\ \frac{3^{17\beta} K \delta}{1 - 3^{17\beta+\alpha} K} \|u\|_{\mathcal{X}}^{\alpha}, & \text{for } \alpha < -17\beta \end{cases}$$

for all $u \in \mathcal{X}$.

Proof By replacing $\psi(u, v) = \delta \|u\|_{\mathcal{X}}^{\frac{\alpha}{2}} \|v\|_{\mathcal{X}}^{\frac{\alpha}{2}}$ and considering $\alpha > -17\beta$ in Theorem 4.2.1 and $\alpha < -17\beta$ in Theorem 4.2.2, we acquire required results. □

The following theorems and corollaries are the various stability results investigated for the multiplicative inverse octadecic functional equation (4.2). Since arguments of proofs are similar to the above theorems and corollaries, we present the statements only.

Theorem 4.2.6 *Let $\psi : \mathcal{X} \times \mathcal{X} \longrightarrow [0, \infty)$ be a mapping satisfying the following inequality:*

$$\sum_{j=0}^{\infty} \left(\frac{K}{3^{18\beta}} \right)^j \psi \left(\frac{u}{3^{j+1}}, \frac{v}{3^{j+1}} \right) < \infty \tag{4.16}$$

for all $u, v \in \mathcal{X}$. Let $m_o : \mathcal{X} \longrightarrow \mathcal{Y}$ be a mapping such that

$$\| D_2 m_o(u, v) \|_{\mathcal{Y}} \leq \psi(u, v) \tag{4.17}$$

for all $u, v \in \mathcal{X}$. Then there exists a unique multiplicative inverse octadecic mapping $\mathcal{M}_O : \mathcal{X} \longrightarrow \mathcal{Y}$ satisfying (4.2) and

$$\| m_o(u) - \mathcal{M}_O(u) \|_Y \leq K \sum_{j=0}^{\infty} \left(\frac{K}{3^{18\beta}} \right)^j \psi \left(\frac{u}{3^{j+1}}, \frac{u}{3^{j+1}} \right) \tag{4.18}$$

for all $u \in \mathcal{X}$. The mapping $R_d(y)$ is defined by

$$\mathcal{M}_O(u) = \lim_{m \to \infty} \frac{1}{3^{18m}} m_o \left(\frac{u}{3^m} \right)$$

for all $u \in \mathcal{X}$.

Theorem 4.2.7 *Let $\psi : \mathcal{X} \times \mathcal{X} \longrightarrow [0, \infty)$ be a mapping satisfying*

$$\sum_{j=0}^{\infty} \left(3^{18\beta} K \right)^j \psi \left(3^j u_1, 3^j v \right) < \infty \tag{4.19}$$

for all $u, v \in \mathcal{X}$. Let $m_s : \mathcal{X} \longrightarrow \mathcal{Y}$ be a mapping satisfying (4.17) for all $u, v \in \mathcal{X}$. Then there exists a unique multiplicative inverse octadecic mapping $\mathcal{M}_O : \mathcal{X} \longrightarrow \mathcal{Y}$ satisfying (4.2) and

$$\| m_o(u) - \mathcal{M}_O(u) \|_{\mathcal{Y}} \leq 3^{18\beta} K \sum_{j=0}^{\infty} \left(3^{18\beta} K \right)^j \psi \left(3^j u, 3^j u \right) \tag{4.20}$$

for all $u \in \mathcal{X}$. The mapping $\mathcal{M}_O(u)$ is defined by

$$\mathcal{M}_O(u) = \lim_{m \to \infty} 3^{18m} m_o \left(3^m u \right)$$

for all $u \in \mathcal{X}$.

Corollary 4.2.8 *Let $\epsilon \geq 0$ be fixed. If a mapping $m_o : \mathcal{X} \longrightarrow \mathcal{Y}$ satisfies the inequality*

$$\| D_2 m_o(u, v) \|_{\mathcal{Y}} \leq \epsilon$$

for all $u, v \in \mathcal{X}$, then there exists a unique multiplicative inverse octadecic mapping $\mathcal{M}_O : \mathcal{X} \longrightarrow \mathcal{Y}$ satisfying (4.2) and

$$\|m_o(u) - \mathcal{M}_O(u)\|_{\mathcal{Y}} \le \frac{3^{18\beta} K \epsilon}{3^{18\beta} - K}$$

for all $u \in \mathcal{X}$.

Corollary 4.2.9 *Let $\delta \ge 0$ be fixed and $\alpha \ne -18\beta$. If a mapping $r : \mathcal{X} \longrightarrow \mathcal{Y}$ satisfies the inequality*

$$\|D_2 m_o(u, v)\|_{\mathcal{Y}} \le \delta \left(\|u\|_{\mathcal{X}}^{\alpha} + \|v\|_{\mathcal{X}}^{\alpha} \right)$$

for all $u, v \in \mathcal{X}$, then there exists a unique multiplicative inverse octadecic mapping $\mathcal{M}_S : \mathcal{X} \longrightarrow \mathcal{Y}$ satisfying (4.2) and

$$\|m_o(u) - \mathcal{M}_O(u)\|_{\mathcal{Y}} \le \begin{cases} \frac{3^{18\beta+1} K \delta}{3^{18\beta+\alpha} - K} \|u\|_{\mathcal{X}}^{\alpha}, & \text{for } \alpha > -18\beta, \\ \frac{3^{18\beta+1} K \delta}{1 - 3^{18\beta+\alpha} K} \|u\|_{\mathcal{X}}^{\alpha}, & \text{for } \alpha < -18\beta \end{cases}$$

for all $u \in \mathcal{X}$.

Corollary 4.2.10 *Let $\delta \ge 0$ be a fixed real number and α be such that $\alpha \ne -18\beta$. If a mapping $m_o : \mathcal{X} \longrightarrow \mathcal{Y}$ satisfies the inequality*

$$\|D_2 m_o(u, v)\|_{\mathcal{Y}} \le \delta \|u\|_{\mathcal{X}}^{\alpha/2} \|v\|_{\mathcal{X}}^{\alpha/2}$$

for all $u, v \in \mathcal{X}$, then there exists a unique multiplicative inverse octadecic mapping $\mathcal{M}_O : \mathcal{X} \longrightarrow \mathcal{Y}$ satisfying (4.2) and

$$\|m_o(u) - \mathcal{M}_O(u)\|_{\mathcal{Y}} \le \begin{cases} \frac{3^{18\beta} K \delta}{3^{18\beta+\alpha} - K} \|u\|_{\mathcal{X}}^{\alpha}, & \text{for } \alpha > -18\beta \\ \frac{3^{18\beta} K \delta}{1 - 3^{18\beta+\alpha} K} \|u\|_{\mathcal{X}}^{\alpha}, & \text{for } \alpha < -18\beta \end{cases}$$

for all $u \in \mathcal{X}$.

4.3　Counter-Examples

In this section, we show that the Eqs. (4.1) and (4.2) are not valid for $\alpha = -17$ in Corollary 4.2.4 and $\alpha = -18$ in Corollary 4.2.9, respectively, in the setting of non-zero real numbers.

Example 4.3.1 Let us define the function

$$\chi(u) = \begin{cases} \frac{c}{u^{17}}, & \text{for } u \in (1, \infty) \\ c, & \text{elsewhere} \end{cases} \tag{4.21}$$

where $\chi : \mathbb{R}^* \longrightarrow \mathbb{R}$. Let $m_s : \mathbb{R}^* \longrightarrow \mathbb{R}$ be a function defined as

$$m_s(u) = \sum_{m=0}^{\infty} 129140163^{-m} \chi(3^{-m} u) \qquad (4.22)$$

for all $u \in \mathbb{R}$. Suppose the mapping $m_s : \mathbb{R}^* \longrightarrow \mathbb{R}$ described in (4.22) satisfies the inequality

$$|D_1 m_s(u, v)| \leq \frac{193710245\, c}{64570081} \left(|u|^{-17} + |v|^{-17}\right) \qquad (4.23)$$

for all $u, v \in \mathbb{R}^*$. We prove that there do not exist a multiplicative inverse septendecic mapping $\mathcal{M}_O : \mathbb{R}^* \longrightarrow \mathbb{R}$ and a constant $\delta > 0$ such that

$$|m_s(u) - \mathcal{M}_O(u)| \leq \delta\, |u|^{-17} \qquad (4.24)$$

for all $u \in \mathbb{R}^*$. Firstly, let us prove that m_s satisfies (4.23). Using (4.21), we have

$$|m_s(u)| = \left| \sum_{m=0}^{\infty} 129140163^{-m} \chi(3^{-m} u) \right| \leq \sum_{m=0}^{\infty} \frac{c}{129140163^m} = \frac{129140163}{129140162} c.$$

We find that that m_s is bounded by $\frac{129140163\, c}{129140162}$ on $\mathbb{R}$. If $|u|^{-17} + |v|^{-17} \geq 1$, then the left hand side of (4.23) is less than $\frac{193710245\, c}{64570081}$. Now, suppose that $0 < |u|^{-17} + |v|^{-17} < 1$. Hence, there exists a positive integer m such that

$$\frac{1}{129140163^{m+1}} \leq |u|^{-17} + |v|^{-17} < \frac{1}{129140163^m}. \qquad (4.25)$$

Hence, the inequality (4.25) produces $129140163^m \left(|u|^{-17} + |v|^{-17}\right) < 1$, or equivalently; $129140163^m u^{-17} < 1$, $129140163^m v^{-17} < 1$. So, $\frac{u^{17}}{129140163^m} > 1$, $\frac{v^{17}}{129140163^m} > 1$. Hence, the last inequalities imply $\frac{u^{17}}{129140163^{m-1}} > 129140163 > 1$, $\frac{v^{17}}{129140163^{m-1}} > 129140163 > 1$ and as a result, we find $\frac{1}{3^{m-1}}(u) > 1$, $\frac{1}{3^{m-1}}(v) > 1$, $\frac{1}{3^{m-1}}(2u + v) > 1$, $\frac{1}{3^{m-1}}(2u - v) > 1$. Hence, for every value of $m = 0, 1, 2, \ldots, n - 1$, we obtain

$$\frac{1}{3^n}(u) > 1, \; \frac{1}{3^n}(v) > 1, \; \frac{1}{3^n}(2u + v) > 1, \; \frac{1}{3^n}(2u - v) > 1$$

and $D_1 m_s(3^{-n} u, 3^{-n} v) = 0$ for $m = 0, 1, 2, \ldots, n - 1$. Applying (4.21) and the definition of m_s, we obtain

$$|D_1 m_s(u, v)| \le \sum_{m=n}^{\infty} \frac{c}{129140163^m} + \sum_{m=n}^{\infty} \frac{c}{129140163^m} + \frac{129140164}{129140163} \sum_{m=n}^{\infty} \frac{c}{129140163^m}$$

$$\le \frac{387420490\, c}{129140163} \frac{1}{129140163^m} \left(1 - \frac{1}{129140163}\right)^{-1}$$

$$\le \frac{387420490\, c}{129140162} \frac{1}{129140163^{m+1}}$$

$$\le \frac{193710245\, c}{64570081} \left(|u|^{-17} + |v|^{-17}\right)$$

for all $u, v \in \mathbb{R}^*$. This means that the inequality (4.23) holds. We claim that the multiplicative inverse septendecic functional equation (4.1) is unstable for $\alpha = -17$ in Corollary 4.2.4. Assume that there exists a multiplicative inverse septendecic mapping $m_s : \mathbb{R}^* \longrightarrow \mathbb{R}$ satisfying (4.24). So, we have

$$|m_s(u)| \le (\delta + 1)|u|^{-17}. \tag{4.26}$$

Moreover, it is possible to choose a positive integer m with the condition $mc > \delta + 1$. If $u \in \left(1, 3^{m-1}\right)$, then $3^{-n}u \in (1, \infty)$ for all $m = 0, 1, 2, \ldots, n - 1$ and thus

$$|m_s(u)| = \sum_{m=0}^{\infty} \frac{\chi(3^{-m}u)}{129140163^m} \ge \sum_{m=0}^{n-1} \frac{\frac{129140163^m c}{u^{17}}}{129140163^m} = \frac{mc}{u^{17}} > (\delta + 1)u^{-17}$$

which contradicts (4.26). Therefore, the multiplicative inverse septendecic functional equation (4.1) is unstable for $\alpha = -17$ in Corollary 4.2.4.

Similar to Example 4.3.1, the following example acts as a counter-example that the equation (4.2) is not stable for $\alpha = -18$ in Corollary 4.2.9.

Example 4.3.2 Define the function $\xi : \mathbb{R}^* \longrightarrow \mathbb{R}$ via

$$\xi(u) = \begin{cases} \frac{\lambda}{u^{18}} & \text{for } u \in (1, \infty) \\ c, & \text{otherwise} \end{cases}. \tag{4.27}$$

Let $m_o : \mathbb{R}^* \longrightarrow \mathbb{R}$ be defined by

$$m_o(u) = \sum_{m=0}^{\infty} 387420489^{-m} \xi(3^{-m}s)$$

for all $u \in \mathbb{R}$. Suppose the function m_o satisfies the inequality

$$|D_2 m_s(u, v)| \le \frac{290565367\, \lambda}{96855122} \left(|u|^{-18} + |v|^{-18}\right)$$

for all $u, v \in \mathbb{R}^*$. Then, there do not exist a multiplicative inverse octadecic mapping $\mathcal{M}_O : \mathbb{R}^* \longrightarrow \mathbb{R}$ and a constant $\eta > 0$ such that

$$|m_s(u) - \mathcal{M}_O(u)| \leq \eta \, |u|^{-18}$$

for all $u \in \mathbb{R}^*$.

Chapter 5
Ulam Stabilities of Multiplicative Inverse Type Novemdecic and Vigintic Functional Equations in Intuitionistic Fuzzy Normed Spaces

Abstract This chapter is devoted to study various classical stability results of multiplicative inverse novemdecic and vigintic functional equations in intuitionistic fuzzy normed spaces and also counter-examples to disprove the validity of stability results for singular cases.

5.1 Introduction

The concept of fuzzy sets was first introduced by Zadeh [150] in 1965 which is a powerful tool for modeling uncertainty and vagueness in various applied problems arising in the field of science and engineering, e.g., population dynamics, chaos control, computer programming, nonlinear dynamical systems, fuzzy physics, nonlinear operators, statistical convergence, etc. For the last four decades, fuzzy theory has become very active area of research and a lot of developments have been made in the theory of fuzzy sets to find the fuzzy analogues of the classical set theory. The fuzzy topology [66] proves to be a very useful tool to deal with such situations where the use of classical theories breaks down.

The concept of intuitionistic fuzzy norm (see [78, 85, 87–89, 99, 126]) is also useful to deal with the inexactness and vagueness arising in modeling.

The generalized Hyers-Ulam stability of various functional equations in intuitionistic fuzzy normed space has been studied in [79, 84, 127, 128, 148]. Saadati et al. [127] introduced the notation of intuitionistic random normed spaces, and then by virtue of this notation to study the stability of a quartic functional equation in the setting of these spaces under arbitrary triangle norms. Mursaleen and Mohiuddine [84] linked two different disciplines, namely, the fuzzy spaces and functional equations. They also proved that the existence of a solution for any approximately cubic mapping implies the completeness of intuitionistic fuzzy normed spaces.

In this chapter, various classical stability results of a multiplicative inverse novemdecic functional equation

B. V. Senthil Kumar and H. Dutta, *Multiplicative Inverse Functional Equations*, Studies in Systems, Decision and Control 289, https://doi.org/10.1007/978-3-030-45355-8_5

57

$$m_n(2x + y) + m_n(2x - y)$$

$$= \frac{4m_n(x)m_n(y)}{\left(4m_n(y)^{2/19} - m_n(x)^{2/19}\right)^{19}} \left[\frac{1}{2}\sum_{k=0}^{9}\binom{19}{2k}[m_n(u)]^{2k/19}[m_n(y)]^{(19-2k)/19}\right]$$

$$(5.1)$$

and a multiplicative inverse vigintic functional equation

$$m_v(2x + y) + m_v(2x - y)$$

$$= \frac{4m_v(x)m_v(y)}{\left(4m_v(y)^{1/10} - m_v(x)^{1/10}\right)^{20}} \left[\sum_{k=0}^{10}\binom{20}{2k}[m_v(x)]^{k/10}[m_v(y)]^{(20-2k)/20}\right] \quad (5.2)$$

are studied in the setting of intuitionistic fuzzy normed spaces. Suitable counter-examples are also presented to disprove the validity of stability results of equations (5.1) and (5.2) for singular cases.

5.2 Preliminaries

In this section, we recall some notations and basic definitions used throughout this chapter.

Definition 5.2.1 A binary operation $* : [0, 1] \times [0, 1] \to [0, 1]$ is said to be a *continuous t-norm* if it satisfies the following conditions:

(i) $*$ is associative and commutative;
(ii) $*$ is continuous;
(iii) $a * 1 = a$ for all $a \in [0, 1]$;
(iv) $a * b \leq c * d$ whenever $a \leq c$ and $b \leq d$ for each $a, b, c, d \in [0, 1]$.

Definition 5.2.2 A binary operation $\Diamond : [0, 1] \times [0, 1] \to [0, 1]$ is said to be a *continuous t-conorm* if it satisfies the following conditions:

(i) $\Diamond$ is associative and commutative;
(ii) $\Diamond$ is continuous;
(iii) $a \Diamond 0 = a$ for all $a \in [0, 1]$;
(iv) $a \Diamond b \leq c \Diamond d$ whenever $a \leq c$ and $b \leq d$ for each $a, b, c, d \in [0, 1]$.

Using the notions of continuous t-norm and t-conorm, Saadati and Park [126] introduced the concept of intuitionistic fuzzy normed space as follows:

Definition 5.2.3 The five-tuple $(X, \mu, \nu, *, \Diamond)$ is said to be an intuitionistic fuzzy normed space (for short, **IFNS**) if X is a vector space, $*$ is a continuous t-norm, $\Diamond$ is a continuous t-conorm, and μ, ν are fuzzy sets on $X \times (0, \infty)$ satisfying the following conditions for each $x, y \in X$ and $s, t > 0$

  (i) $\mu(x, t) + \nu(x, t) \leq 1$;

 (ii) $\mu(x, t) > 0$;

(iii) $\mu(x, t) = 1$ if and only if $x = 0$;

(iv) $\mu(\alpha x, t) = \mu\left(x, \frac{t}{|\alpha|}\right)$ for each $\alpha \neq 0$;

 (v) $\mu(x, t) * \mu(y, s) \leq \mu(x + y, t + s)$;

(vi) $\mu(x, \cdot) : (0, \infty) \to [0, 1]$ is continuous;

(vii) $\lim_{t \to \infty} \mu(x, t) = 1$ and $\lim_{t \to 0} \mu(x, t) = 0$;

(viii) $\nu(x, t) < 1$;

 (ix) $\nu(x, t) = 0$ if and only if $x = 0$;

 (x) $\nu(\alpha x, t) = \nu\left(x, \frac{t}{|\alpha|}\right)$ for each $\alpha \neq 0$;

 (xi) $\nu(x, t) \Diamond \nu(y, s) \geq \nu(x + y, t + s)$;

(xii) $\nu(x, \cdot) : (0, \infty) \to [0, 1]$ is continuous;

(xiii) $\lim_{t \to \infty} \nu(x, t) = 0$ and $\lim_{t \to 0} \nu(x, t) = 1$.

In this case (μ, ν) is called an intuitionistic fuzzy norm.

Example 5.2.4 Let $(X, \|\cdot\|)$ be a normed space, $a * b = ab$ and $a \Diamond b = \min\{a + b, 1\}$ for all $a, b \in [0, 1]$. For all $x \in X$ and every $t > 0$, consider

$$\mu(x, t) = \begin{cases} \frac{t}{t + \|x\|} & \text{if } t > 0 \\ 0 & \text{if } t \leq 0; \end{cases} \quad \text{and} \quad \nu(x, t) = \begin{cases} \frac{\|x\|}{t + \|x\|} & \text{if } t > 0 \\ 0 & \text{if } t \leq 0; \end{cases}$$

Then $(X, \mu, \nu, *, \Diamond)$ is an IFNS.

The concepts of convergence and Cauchy sequence in intuitionistic fuzzy normed space are studied in [126].

Let $(X, \mu, \nu, *, \Diamond)$ be an IFNS. A sequence $x = (x_k)$ is said to be intuitionistic fuzzy convergent to $L \in X$ if, for every $\epsilon > 0$, there exists $k_0 \in \mathbb{N}$ such that $\mu(x_k - L, t) > 1 - \epsilon$ and $\nu(x_k - L, t) < \epsilon$ for all $k \geq k_0$. In this case, we write $(\mu, \nu) - \lim x_k = L$ or $x_k \xrightarrow{(\mu, \nu)} L$ as $k \to \infty$.

Let $(X, \mu, \nu, *, \Diamond)$ be an IFNS. A sequence $x = (x_k)$ is said to be intuitionistic fuzzy Cauchy sequence if, for every $\epsilon > 0$ and $t > 0$, there exists $k_0 \in \mathbb{N}$ such that $\mu(x_k - x_l, t) > 1 - \epsilon$ and $\nu(x_k - x_l, t) < \epsilon$ for all $k, l \geq k_0$.

An IFNS $(X, \mu, \nu, *, \Diamond)$ is said to be complete if every intuitionistic fuzzy Cauchy sequence is intuitionistic fuzzy convergent in $(X, \mu, \nu, *, \Diamond)$. In this case (X, μ, ν) is called intuitionistic fuzzy Banach space.

5.3 Ulam Stabilities of Eqs. (5.1) and (5.2)

Throughout this section, let us assume that X to be a linear space and (Y, μ, ν) an intuitionistic fuzzy Banach space. For the sake of convenience, we denote for given mappings $m_N, m_V : X \to Y$, the difference operators $D_1 m_N, D_2 m_V : X \times X \to Y$ by

$$D_1 M_N(x, y)$$
$$= M_N(2x + y) + M_N(2x - y)$$
$$- \frac{4M_N(x)M_N(y)}{\left(4M_N(y)^{2/19} - M_N(x)^{2/19}\right)^{19}} \left[\frac{1}{2} \sum_{k=0}^{9} \binom{19}{2k} [M_N(u)]^{2k/19} [M_N(y)]^{(19-2k)/19}\right]$$

and

$$D_2 M_V(x, y)$$
$$= M_V(2x + y) + M_V(2x - y)$$
$$- \frac{4M_V(x)M_V(y)}{\left(4M_V(y)^{1/10} - M_V(x)^{1/10}\right)^{20}} \left[\sum_{k=0}^{10} \binom{20}{2k} [M_V(x)]^{k/10} [M_V(y)]^{(20-2k)/20}\right]$$

for all $x, y \in X$.

Theorem 5.3.1 *Let $\phi : X \times X \longrightarrow [0, \infty)$ be a function such that*

$$\Psi(x, y) = \sum_{n=0}^{\infty} 3^{19n} \phi\left(3^n x, 3^n y\right) < \infty \tag{5.3}$$

for all $x, y \in X$. Let $M_N : X \longrightarrow Y$ be a function such that

$$\left.\begin{array}{l} \lim_{t \to \infty} \mu\left(M_N(x, y), t\phi(x, y)\right) = 1 \\ \lim_{t \to \infty} \nu\left(M_N(x, y), t\phi(x, y)\right) = 0 \end{array}\right\} \tag{5.4}$$

uniformly in $X \times X$. Then $\mathcal{M}_n(x) = (\mu, \nu) - \lim_{n \to \infty} 3^{19n} M_N\left(3^n x\right)$ for each $x \in X$ exists and defines a multiplicative inverse novemdecic mapping $\mathcal{M}_n : X \longrightarrow Y$ such that if for some $\delta > 0$, $\alpha > 0$ and all $x, y \in X$,

$$\left.\begin{array}{l} \mu\left(D_1 M_N(x, y), \delta\phi(x, y)\right) > \alpha \\ \nu\left(D_1 M_N(x, y), \delta\phi(x, y)\right) < 1 - \alpha \end{array}\right\} \tag{5.5}$$

then

$$\left.\begin{array}{l} \mu\left(\mathcal{M}_n(x) - M_N(x), \dfrac{3^{19}\delta}{2}\Psi(x, x)\right) > \alpha \\ \nu\left(\mathcal{M}_n(x) - M_N(x), \dfrac{3^{19}\delta}{2}\Psi(x, x)\right) < 1 - \alpha. \end{array}\right\}$$

Also, the multiplicative inverse novemdecic mapping $\mathcal{M}_n$ is unique such that

$$\left.\begin{array}{l} \lim_{n \to \infty} \mu \left(\mathcal{M}_n(x) - M_N(x), \frac{3^{19}t}{2} \Psi(x, x) \right) = 1 \\[2mm] \lim_{n \to \infty} \nu \left(\mathcal{M}_n(x) - M_N(x), \frac{3^{19}t}{2} \Psi(x, x) \right) = 0 \end{array}\right\} \qquad (5.6)$$

uniformly in X.

Proof Let $\epsilon > 0$ be given. Using (5.4), we can find some $t_0 > 0$ such that

$$\left.\begin{array}{l} \mu \left(D_1 M_N(x, y), t\phi(x, y) \right) \geq 1 - \epsilon \\[2mm] \nu \left(D_1 M_N(x, y), t\phi(x, y) \right) \leq \epsilon \end{array}\right\} \qquad (5.7)$$

for all $x, y \in X$ and all $t \geq t_0$. Substituting $y = x$ in (5.7), we obtain

$$\left.\begin{array}{l} \mu \left(3^{19} M_N(3x) - M_N(x), \frac{3^{19}t}{2} \phi(x, x) \right) \geq 1 - \epsilon \\[2mm] \nu \left(3^{19} M_N(3x) - M_N(x), \frac{3^{19}t}{2} \phi(x, x) \right) \leq \epsilon \end{array}\right\} \qquad (5.8)$$

for all $x, y \in X$ and all $t \geq t_0$. Now, replacing x by $3x$ in (5.8), we get

$$\left.\begin{array}{l} \mu \left(3^{38} M_N \left(3^2 x \right) - 3^{19} M_N(3x), \frac{3^{38}t}{2} \phi(3x, 3x) \right) \geq 1 - \epsilon \\[2mm] \nu \left(3^{38} M_N \left(3^2 x \right) - 3^{19} M_N(3x), \frac{3^{38}t}{2} \phi(3x, 3x) \right) \leq \epsilon \end{array}\right\} \qquad (5.9)$$

for all $x, y \in X$ and all $t \geq t_0$. Combining (5.8) and (5.9) yields,

$$\mu \left(3^{38} M_N \left(3^2 x \right) - M_N(x), \frac{3^{19}t}{2} \sum_{k=0}^{1} 3^{19k} \phi \left(3^k x, 3^k x \right) \right)$$

$$\geq \mu \left(3^{38} M_N \left(3^2 x \right) - 3^{19} M_N(3x), \frac{3^{19}t}{2} \phi(3x, 3x) \right)$$

$$* \mu \left(3^{19} M_N(3x) - M_N(x), \frac{3^{19}t}{2} \phi(x, x) \right)$$

$$\geq (1 - \epsilon) * (1 - \epsilon) = 1 - \epsilon$$

and

$$\nu \left(3^{38} M_N \left(3^2 x \right) - M_N(x), \frac{3^{19} t}{2} \sum_{k=0}^{1} 3^{18k} \phi \left(3^k x, 3^k x \right) \right)$$

$$\leq \nu \left(3^{38} M_N \left(3^2 x \right) - 3^{19} M_N(3x), \frac{3^{19} t}{2} \phi(3x, 3x) \right)$$

$$\Diamond \nu \left(3^{19} M_N(3x) - M_N(x), \frac{3^{19} t}{2} \phi(x, x) \right)$$

$$\leq \epsilon \Diamond \epsilon = \epsilon$$

for all $x, y \in X$ and all $t \geq t_0$. Proceeding further and using induction on a positive integer n, we get

$$\left. \begin{aligned} \mu \left(3^{19n} M_N \left(3^n x \right) - M_N(x), \frac{3^{19} t}{2} \sum_{k=0}^{n-1} 3^{19k} \phi \left(3^k x, 3^k x \right) \right) &\geq 1 - \epsilon \\ \nu \left(3^{19n} M_N \left(3^n x \right) - M_N(x), \frac{3^{19} t}{2} \sum_{k=0}^{n-1} 3^{19k} \phi \left(3^k x, 3^k x \right) \right) &\leq \epsilon \end{aligned} \right\} \tag{5.10}$$

for all $x, y \in X$ and all $t \geq t_0$. In order to prove the convergence of the sequence $\left\{ 3^{19n} M_N \left(3^n x \right) \right\}$, letting $t = t_0$ and replacing (x, y) by $(3^m x, 3^m y)$ in (5.10), we find that for $n > m > 0$

$$\left. \begin{aligned} \mu \Big(3^{19(n+m)} M_N \left(3^{n+m} x \right) &- 3^{19m} M_N \left(3^m x \right), \\ \frac{3^{19} t_0}{2} \sum_{k=0}^{n-1} 3^{19(k+m)} &\phi \left(3^{k+m} x, 3^{k+m} x \right) \Big) \\ &\geq 1 - \epsilon \\ \nu \Big(3^{19(n+m)} M_N \left(3^{n+m} x \right) &- 3^{19m} M_N \left(3^m x \right), \\ \frac{3^{19} t_0}{2} \sum_{k=0}^{n-1} 3^{19(k+m)} &\phi \left(3^{k+m} x, 3^{k+m} x \right) \Big) \\ &\leq \epsilon. \end{aligned} \right\} \tag{5.11}$$

The convergence of (5.3) and

$$\frac{3^{19}}{2} \sum_{k=0}^{n-1} 3^{19(m+k)} \phi \left(3^{m+k} x, 3^{m+k} x \right) = \frac{3^{19}}{2} \sum_{k=m}^{m+n-1} 3^{19k} \phi \left(3^k x, 3^k x \right)$$

imply that for given $\delta > 0$ there is $n_0 \in \mathbb{N}$ such that

$$\frac{3^{19}t_0}{2}\sum_{k=m}^{m+n-1}3^{19k}\phi\left(3^kx,3^kx\right)<\delta,$$

for all $m \geq n_0$ and all $n > 0$. From (5.11), we deduce that

$$\mu\left(3^{19(m+n)}M_N\left(3^{m+n}x\right)-3^{19m}M_N\left(3^mx\right),\delta\right)$$
$$\geq\mu\left(3^{19(m+n)}M_N\left(3^{m+n}x\right)-3^{19m}M_N\left(3^mx\right),\right.$$
$$\left.\frac{3^{19}t_0}{2}\sum_{k=0}^{n-1}3^{19(m+k)}\phi\left(3^{m+k}x,3^{m+k}x\right)\right)\geq1-\epsilon$$

and

$$\nu\left(3^{19(m+n)}M_N\left(3^{m+n}x\right)-3^{19m}M_N\left(3^mx\right),\delta\right)$$
$$\leq\nu\left(3^{19(m+n)}M_N\left(3^{m+n}x\right)-3^{19m}M_N\left(3^mx\right),\right.$$
$$\left.\frac{3^{19}t_0}{2}\sum_{k=0}^{n-1}3^{19(m+k)}\phi\left(3^{m+k}x,3^{m+k}x\right)\right)\leq\epsilon$$

for all $m \geq n_0$ and all $n > 0$. Hence $\left\{3^{19n}M_N\left(3^nx\right)\right\}$ is a Cauchy sequence in Y. Since Y is an intuitionistic fuzzy Banach space, the sequence $\left\{3^{19n}M_N\left(3^nx\right)\right\}$ converges to some $\mathcal{M}_n \in Y$. Hence we can define a mapping $\mathcal{M}_n : X \longrightarrow Y$ such that $\mathcal{M}_n(x) = (\mu,\nu)-\lim\limits_{n\to\infty}3^{19n}M_N\left(3^nx\right)$, namely, for each $t > 0$, and $x \in X$,

$$\mu\left(\mathcal{M}_n(x)-3^{19n}M_N\left(3^nx\right),t\right)=1\quad\text{and}\quad\nu\left(\mathcal{M}_n(x)-3^{19n}M_N\left(3^nx\right),t\right)=0.$$

Taking the limit $n \to \infty$ in (5.10), we see that the existence of (5.6) uniformly in X. Now, let $x, y \in X$. Choose any fixed value of $t > 0$, and $\epsilon \in (0, 1)$. Since $\lim\limits_{n\to\infty}3^{19n}\phi\left(3^nx,3^ny\right)=0$, there exists $n_1 \geq n_0$ such that $t_0\phi\left(3^nx,3^ny\right)<\frac{t}{4\cdot3^{19n}}$ for all $n \geq n_1$. Hence for each $n \geq n_1$, we have

$$\mu\left(D_1\mathcal{M}_n(x,y),t\right)$$
$$\geq\mu\left(\mathcal{M}_n(2x+y)-3^{19n}M_N\left(3^n(2x+y)\right),\frac{t}{4}\right)$$
$$*\mu\left(\mathcal{M}_n(2x-y)-3^{19n}M_N\left(3^n(2x-y)\right),\frac{t}{4}\right)$$
$$*\mu\left(\frac{4\mathcal{M}_n(x)\mathcal{M}_n(y)}{\left(4\mathcal{M}_n(y)^{2/19}-\mathcal{M}_n(x)^{2/19}\right)^{19}}\left[\frac{1}{2}\sum_{k=0}^{9}\binom{19}{2k}[\mathcal{M}_n(u)]^{2k/19}[\mathcal{M}_n(y)]^{(19-2k)/19}\right]\right.$$
$$\left.-\frac{4\cdot3^{19n}M_N(3^nx)3^{19n}M_N(3^ny)}{\left(4\cdot3^{19n}M_N(3^ny)^{2/19}-3^{19n}M_N(3^nx)^{2/19}\right)^{19}}\right.$$

$$\left[\frac{1}{2}\sum_{k=0}^{9}\binom{19}{2k}[3^{19n}M_N(3^nx)]^{2k/19}[3^{19n}M_N(3^ny)]^{(19-2k)/19}\right],\frac{t}{4}\right)$$

$$* \mu\left(D_1M_N\left(3^nx,3^ny\right),\frac{t}{4\cdot3^{19n}}\right)$$

(5.12)

and also

$$\mu\left(D_1M_N\left(3^nx,3^ny\right),\frac{t}{4\cdot3^{19n}}\right)\geq\mu\left(D_1M_N\left(3^nx,3^ny\right),t_0\phi\left(3^nx,3^ny\right)\right).$$

(5.13)

Letting $n\to\infty$ in (5.12) and using (5.7), (5.13), we get

$$\mu\left(D_1\mathcal{M}_n(x,y),t\right)\geq 1-\epsilon$$

for all $t>0$ and $\epsilon\in(0,1)$. Similarly, we obtain

$$\nu\left(D_1\mathcal{M}_n(x,y),t\right)\leq\epsilon$$

for all $t>0$ and $\epsilon\in(0,1)$. It follows that

$$\mu\left(D_1\mathcal{M}_n(x,y),t\right)=1\quad\text{and}\quad\nu\left(D_1\mathcal{M}_n(x,y),t\right)=0,$$

for all $t>0$. Therefore $\mathcal{M}_n$ satisfies (5.1), which shows that $\mathcal{M}_n$ is multiplicative inverse novemdecic mapping. Next, suppose that for some positive δ and α, (5.5) holds and

$$\phi_n(x,y)=\frac{1}{2}\sum_{k=0}^{n-1}3^{19k}\phi\left(3^kx,3^ky\right),$$

for all $x,y\in X$. By similar argument as in the beginning of the proof we can deduce from (5.5)

$$\left.\begin{array}{l}\mu\left(3^{19n}M_N\left(3^nx\right)-M_N(x),\frac{3^{19}\delta}{4}\sum_{k=0}^{n-1}3^{19k}\phi\left(3^kx,3^ky\right)\right)\geq\alpha\\[4mm]\nu\left(3^{19n}M_N\left(3^nx\right)-M_N(x),\frac{3^{19}\delta}{4}\sum_{k=0}^{n-1}3^{19k}\phi\left(3^kx,3^ky\right)\right)\leq 1-\alpha,\end{array}\right\}$$

(5.14)

for all positive integers n. For $s>0$, we have

$$\left.\begin{aligned}
\mu \left(\mathcal{M}_n(x) - M_N(x), \delta\phi_n(x, x) + s \right) \qquad\qquad\qquad \\
\geq \mu \left(3^{19n} M_N \left(3^n x \right) - M_N(x), \delta\phi_n(x, x) \right) \\
* \, \mu \left(\mathcal{M}_n(x) - 3^{19n} M_N \left(3^n x \right), s \right) \\
\nu \left(\mathcal{M}_n(x) - M_N(x), \delta\phi_n(x, x) + s \right) \qquad\qquad\qquad \\
\leq \nu \left(\mathcal{M}_n(x) - 3^{19n} M_N \left(3^n x \right), s \right) \\
\Diamond \nu \left(3^{19n} M_N \left(3^n x \right) - M_N(x), \delta\phi_n(x, x) \right).
\end{aligned}\right\} \tag{5.15}$$

Combining (5.14), (5.15) and using the fact that

$$\left.\begin{aligned}
\lim_{n \to \infty} \mu \left(\mathcal{M}_n(x) - 3^{19n} M_N \left(3^n x \right), s \right) = 1 \\
\lim_{n \to \infty} \nu \left(\mathcal{M}_n(x) - 3^{19n} M_N \left(3^n x \right), s \right) = 0,
\end{aligned}\right\}$$

we obtain

$$\left.\begin{aligned}
\mu \left(\mathcal{M}_n(x) - M_N(x), \delta\phi_n(x, x) + s \right) \geq \alpha \\
\nu \left(\mathcal{M}_n(x) - M_N(x), \delta\phi_n(x, x) + s \right) \leq 1 - \alpha,
\end{aligned}\right\}$$

for sufficiently large n. From the (upper-semi) continuity of real functions $\mu(\mathcal{M}_n(x) - M_N(x), \cdot)$ and $\nu(\mathcal{M}_n(x) - M_N(x), \cdot)$, we see that

$$\left.\begin{aligned}
\mu \left(\mathcal{M}_n(x) - M_N(x), \frac{3^{19}\delta}{4} \Psi(x, x) + s \right) \geq \alpha \\
\nu \left(\mathcal{M}_n(x) - M_N(x), \frac{3^{19}\delta}{4} \Psi(x, x) + s \right) \leq 1 - \alpha.
\end{aligned}\right\}$$

Taking the limit $s \to \infty$, we get

$$\left.\begin{aligned}
\mu \left(\mathcal{M}_n(x) - M_N(x), \frac{3^{19}\delta}{4} \Psi(x, x) \right) \geq \alpha \\
\nu \left(\mathcal{M}_n(x) - M_N(x), \frac{3^{19}\delta}{4} \Psi(x, x) \right) \leq 1 - \alpha.
\end{aligned}\right\}$$

It remains to prove the uniqueness of $\mathcal{M}_n$. Let $\mathcal{M}'_n$ be another multiplicative inverse novemdecic mapping satisfying (5.6). Choose any fixed value of $c > 0$. Given $\epsilon > 0$, there is some $t_0 > 0$ such that (5.6) for $\mathcal{M}_n$ and $\mathcal{M}'_n$

$$\left.\begin{array}{l} \mu\left(\mathcal{M}_n(x) - M_N(x), \dfrac{3^{19}t}{4}\Psi(x,x)\right) \geq 1 - \epsilon, \\[4mm] \mu\left(\mathcal{M}'_n(x) - M_N(x), \dfrac{3^{19}t}{4}\Psi(x,x)\right) \geq 1 - \epsilon \\[4mm] \nu\left(\mathcal{M}_n(x) - M_N(x), \dfrac{3^{19}t}{4}\Psi(x,x)\right) \leq \epsilon, \\[4mm] \nu\left(\mathcal{M}'_n(x) - M_N(x), \dfrac{3^{19}t}{4}\Psi(x,x)\right) \leq \epsilon \end{array}\right\}$$

for all $x \in X$ and all $t \geq t_0$. For some $x \in X$, we can find some integer n_0 such that

$$t_0 \sum_{k=n}^{\infty} 3^{19k}\phi\left(3^k x, 3^k x\right) < \frac{c}{2}, \qquad \text{for all } n \geq n_0.$$

Since

$$\sum_{k=n}^{\infty} 3^{19k}\phi\left(3^k x, 3^k x\right) = 3^{19n} \sum_{k=n}^{\infty} 3^{19(k-n)}\phi\left(3^{k-n}\left(3^n x\right), 3^{k-n}\left(3^n x\right)\right)$$

$$= 3^{19n} \sum_{m=0}^{\infty} 3^{19m}\phi\left(3^m\left(3^n x\right), 3^m\left(3^n x\right)\right) = 3^{19n}\Psi\left(3^n x, 3^n x\right),$$

we have

$$\mu\left(\mathcal{M}_n(x) - \mathcal{M}'_n(x), c\right)$$

$$\geq \mu\left(\mathcal{M}_n(x) - 3^{19n} M_N\left(3^n x\right), \frac{c}{2}\right) * \mu\left(3^{19n} M_N\left(3^n x\right) - \mathcal{M}'_n(x), \frac{c}{2}\right)$$

$$\geq \mu\left(\mathcal{M}_n\left(3^n x\right) - M_N\left(3^n x\right), \frac{c}{2 \cdot 3^{19n}}\right) * \mu\left(M_N\left(3^n x\right) - \mathcal{M}'_n\left(3^n x\right), \frac{c}{2 \cdot 3^{19n}}\right)$$

$$\geq \mu\left(\mathcal{M}_n\left(3^n x\right) - M_N\left(3^n x\right), \frac{t_0}{3^{19n}} \sum_{k=n}^{\infty} 3^{19k}\phi\left(3^k x, 3^k x\right)\right)$$

$$* \mu\left(M_N\left(3^n x\right) - \mathcal{M}'_n\left(3^n x\right), \frac{t_0}{3^{19n}} \sum_{k=n}^{\infty} 3^{19k}\phi\left(3^k x, 3^k x\right)\right)$$

$$\geq \mu\left(\mathcal{M}_n\left(3^n x\right) - M_N\left(3^n x\right), t_0\Psi\left(3^n x, 3^n x\right)\right)$$

$$* \mu\left(M_N\left(3^n x\right) - \mathcal{M}'_n\left(3^n x\right), t_0\Psi\left(3^n x, 3^n x\right)\right) \geq 1 - \epsilon$$

and similarly

$$\nu\left(\mathcal{M}_n(x) - \mathcal{M}'_n(x), c\right)$$

$$\leq \nu\left(\mathcal{M}_n(x) - 3^{19n} M_N\left(3^n x\right), \frac{c}{2}\right) \Diamond \nu\left(3^{19n} M_N\left(3^n x\right) - \mathcal{M}'_n(x), \frac{c}{2}\right)$$

$$\leq \nu\left(\mathcal{M}_n\left(3^n x\right) - M_N\left(3^n x\right), \frac{c}{2 \cdot 3^{19n}}\right) \Diamond \nu\left(M_N\left(3^n x\right) - \mathcal{M}'_n\left(3^n x\right), \frac{c}{2 \cdot 3^{19n}}\right)$$

$$\leq \nu\left(\mathcal{M}_n\left(3^n x\right) - M_N\left(3^n x\right), \frac{t_0}{3^{19n}} \sum_{k=n}^{\infty} 3^{19k} \phi\left(3^k x, 3^k x\right)\right)$$

$$\Diamond \nu\left(M_N\left(3^n x\right) - \mathcal{M}'_n\left(3^n x\right), \frac{t_0}{3^{19n}} \sum_{k=n}^{\infty} 3^{19k} \phi\left(3^k x, 3^k x\right)\right)$$

$$\leq \nu\left(\mathcal{M}_n\left(3^n x\right) - M_N\left(3^n x\right), t_0 \Psi\left(3^n x, 3^n x\right)\right)$$
$$\Diamond \nu\left(M_N\left(3^n x\right) - \mathcal{M}'_n\left(3^n x\right), t_0 \Psi\left(3^n x, 3^n x\right)\right) \leq \epsilon.$$

It follows that

$$\mu\left(\mathcal{M}_n(x) - \mathcal{M}'_n(x), c\right) = 1 \quad \text{and} \quad \nu\left(\mathcal{M}_n(x) - \mathcal{M}'_n(x), c\right) = 0$$

for all $c > 0$. Hence $\mathcal{M}_n(x) = \mathcal{M}'_n(x)$ for all $x \in X$, which completes the proof of the theorem. $\square$

Corollary 5.3.2 *Let $M_N : X \longrightarrow Y$ be a function such that for all $c_1 \geq 0$, $p < -19$*

$$\left.\begin{array}{l} \lim_{t \to \infty} \mu\left(D_1 M_N(x, y), tc_1\left(\|x\|^p + \|y\|^p\right)\right) = 1 \\ \lim_{t \to \infty} \nu\left(D_1 M_N(x, y), tc_1\left(\|x\|^p + \|y\|^p\right)\right) = 0, \end{array}\right\}$$

uniformly in $X \times X$. Then there exists a unique multiplicative inverse novemdecic mapping $\mathcal{M}_n : X \longrightarrow Y$ satisfying (5.1) such that

$$\left.\begin{array}{l} \lim_{t \to \infty} \mu\left(\mathcal{M}_n(x) - M_N(x), \frac{3^{19} c_1 t \|x\|^p}{1 - 3^{p+19}}\right) = 1 \\ \lim_{t \to \infty} \nu\left(\mathcal{M}_n(x) - M_N(x), \frac{3^{19} c_1 t \|x\|^p}{1 - 3^{p+19}}\right) = 0, \end{array}\right\}$$

uniformly in X.

Proof The proof is obtained by considering $\phi(x, y) = c_1\left(\|x\|^p + \|y\|^p\right)$ for all $x, y \in X$ in Theorem 5.3.1. $\square$

Corollary 5.3.3 *Let $M_N : X \longrightarrow Y$ be a function and suppose that there exist real numbers a, b such that $\rho = a + b < -19$. If there exists $c_2 \geq 0$ such that*

$$\left.\begin{array}{l} \lim_{t \to \infty} \mu\left(D_1 M_N(x, y), tc_2 \|x\|^a \|y\|^b\right) = 1 \\ \lim_{t \to \infty} \nu\left(D_1 M_N(x, y), tc_2 \|x\|^a \|y\|^b\right) = 0, \end{array}\right\}$$

uniformly in $X \times X$. Then there exists a unique multiplicative inverse novemdecic mapping $\mathcal{M}_n : X \longrightarrow Y$ satisfying (5.1) such that

$$\lim_{t \to \infty} \mu \left(\mathcal{M}_n(x) - M_N(x), \frac{3^{19} c_2 t \, \|x\|^{\rho}}{2 \left(1 - 3^{\rho+19}\right)} \right) = 1$$

$$\lim_{t \to \infty} \nu \left(\mathcal{M}_n(x) - M_N(x), \frac{3^{19} c_2 t \, \|x\|^{\rho}}{2 \left(1 - 3^{\rho+19}\right)} \right) = 0,$$

uniformly in X.

Proof It is easy to prove the required results in the Corollary by taking $\phi(x, y) = c_2 \|x\|^a \|y\|^b$, for all $x, y \in X$ in Theorem 5.3.1. □

Corollary 5.3.4 *Let $c_3 \geq 0$ and $\alpha < -19$ be real numbers, and $M_N : X \longrightarrow Y$ be a function such that*

$$\lim_{t \to \infty} \mu \left(D_1 M_N(x, y), t c_3 \left(\|x\|^{\alpha/2} \|y\|^{\alpha/2} + (\|x\|^{\alpha} + \|y\|^{\alpha}) \right) \right) = 1$$

$$\lim_{t \to \infty} \nu \left(D_1 M_N(x, y), t c_3 \left(\|x\|^{\alpha/2} \|y\|^{\alpha/2} + (\|x\|^{\alpha} + \|y\|^{\alpha}) \right) \right) = 0,$$

uniformly in $X \times X$. Then there exists a unique multiplicative inverse novemdecic mapping $\mathcal{M}_n : X \longrightarrow Y$ satisfying (5.1) such that

$$\lim_{t \to \infty} \mu \left(\mathcal{M}_n(x) - M_N(x), \frac{3 \cdot 3^{19} c_3 t \, \|x\|^{\alpha}}{2 \left(1 - 3^{\alpha+19}\right)} \right) = 1$$

$$\lim_{t \to \infty} \nu \left(\mathcal{M}_n(x) - M_N(x), \frac{3 \cdot 3^{19} c_3 t \, \|x\|^{\alpha}}{2 \left(1 - 3^{\alpha+19}\right)} \right) = 0,$$

uniformly in X.

Proof The proof is analogous to the proof of Theorem 5.3.1, by choosing $\phi(x, y) = c_3 \left(\|x\|^{\alpha/2} \|y\|^{\alpha/2} + (\|x\|^{\alpha} + \|y\|^{\alpha}) \right)$, for all $x, y \in X$. □

The ensuing outcomes contain various stability results of equation (5.2) in the setting of intuitionistic fuzzy normed spaces. The stability results of equation (5.2) can be proved by similar arguments as in the stability results of equation (5.1). For the sake of completeness, we will present the statements only.

Theorem 5.3.5 *Let $\phi : X \times X \longrightarrow [0, \infty)$ be a function such that*

$$\Psi(x, y) = \sum_{n=0}^{\infty} 3^{20n} \phi \left(3^n x, 3^n y \right) < \infty \tag{5.16}$$

for all $x, y \in X$. Let $M_V : X \longrightarrow Y$ be a function such that

$$\lim_{t \to \infty} \mu \left(M_V(x, y), t\phi(x, y) \right) = 1$$

$$\lim_{t \to \infty} \nu \left(M_V(x, y), t\phi(x, y) \right) = 0 \tag{5.17}$$

uniformly in $X \times X$. *Then* $\mathcal{M}_v(x) = (\mu, \nu) - \lim\limits_{n \to \infty} 3^{20n} M_V(3^n x)$ *for each* $x \in X$
exists and defines a multiplicative inverse vigintic mapping $\mathcal{M}_v : X \longrightarrow Y$ *such that if for some* $\delta > 0$, $\alpha > 0$ *and all* $x, y \in X$,

$$\left.\begin{aligned} \mu\left(D_2 M_V(x, y), \delta\phi(x, y)\right) > \alpha \\ \nu\left(D_2 M_V(x, y), \delta\phi(x, y)\right) < 1 - \alpha \end{aligned}\right\} \tag{5.18}$$

then

$$\left.\begin{aligned} \mu\left(\mathcal{M}_v(x) - M_V(x), \frac{3^{20}\delta}{2}\Psi(x, x)\right) > \alpha \\ \nu\left(\mathcal{M}_v(x) - M_V(x), \frac{3^{20}\delta}{2}\Psi(x, x)\right) < 1 - \alpha. \end{aligned}\right\}$$

Also, the multiplicative inverse vigintic mapping $\mathcal{M}_v$ *is unique such that*

$$\left.\begin{aligned} \lim_{n \to \infty} \mu\left(\mathcal{M}_v(x) - M_V(x), \frac{3^{20}t}{2}\Psi(x, x)\right) = 1 \\ \lim_{n \to \infty} \nu\left(\mathcal{M}_v(x) - M_V(x), \frac{3^{20}t}{2}\Psi(x, x)\right) = 0 \end{aligned}\right\} \tag{5.19}$$

uniformly in X.

Corollary 5.3.6 *Let* $M_V : X \longrightarrow Y$ *be a function such that for all* $c_1 \geq 0$, $p < -20$

$$\left.\begin{aligned} \lim_{t \to \infty} \mu\left(D_2 M_V(x, y), tc_1\left(\|x\|^p + \|y\|^p\right)\right) = 1 \\ \lim_{t \to \infty} \nu\left(D_2 M_V(x, y), tc_1\left(\|x\|^p + \|y\|^p\right)\right) = 0, \end{aligned}\right\}$$

uniformly in $X \times X$. *Then there exists a unique multiplicative inverse vigintic mapping* $\mathcal{M}_v : X \longrightarrow Y$ *satisfying* (5.2) *such that*

$$\left.\begin{aligned} \lim_{t \to \infty} \mu\left(\mathcal{M}_v(x) - M_V(x), \frac{3^{20}c_1 t\,\|x\|^p}{1 - 3^{p+20}}\right) = 1 \\ \lim_{t \to \infty} \nu\left(\mathcal{M}_v(x) - M_V(x), \frac{3^{20}c_1 t\,\|x\|^p}{1 - 3^{p+20}}\right) = 0, \end{aligned}\right\}$$

uniformly in X.

Corollary 5.3.7 *Let* $M_V : X \longrightarrow Y$ *be a function and suppose that there exist real numbers* a, b *such that* $\rho = a + b < -20$. *If there exists* $c_2 \geq 0$ *such that*

$$\left.\begin{aligned} \lim_{t \to \infty} \mu\left(D_2 M_V(x, y), tc_2\,\|x\|^a\,\|y\|^b\right) = 1 \\ \lim_{t \to \infty} \nu\left(D_2 M_V(x, y), tc_1\,\|x\|^a\,\|y\|^b\right) = 0, \end{aligned}\right\}$$

uniformly in $X \times X$. Then there exists a unique multiplicative inverse vigintic map-
ping $\mathcal{M}_v : X \longrightarrow Y$ satisfying (5.2) such that

$$
\left.
\begin{aligned}
\lim_{t \to \infty} \mu \left(\mathcal{M}_v(x) - M_V(x), \frac{3^{20} c_2 t \, \|x\|^\rho}{2 \left(1 - 3^{\rho+20} \right)} \right) &= 1 \\
\lim_{t \to \infty} \nu \left(\mathcal{M}_v(x) - M_V(x), \frac{3^{20} c_2 t \, \|x\|^\rho}{2 \left(1 - 3^{\rho+20} \right)} \right) &= 0,
\end{aligned}
\right\}
$$

uniformly in X.

Corollary 5.3.8 *Let $c_3 \geq 0$ and $\alpha < -20$ be real numbers, and $M_V : X \longrightarrow Y$ be
a function such that*

$$
\left.
\begin{aligned}
\lim_{t \to \infty} \mu \left(D_2 M_V(x, y), t c_3 \left(\|x\|^{\alpha/2} \|y\|^{\alpha/2} + (\|x\|^\alpha + \|y\|^\alpha) \right) \right) &= 1 \\
\lim_{t \to \infty} \nu \left(D_2 M_V(x, y), t c_3 \left(\|x\|^{\alpha/2} \|y\|^{\alpha/2} + (\|x\|^\alpha + \|y\|^\alpha) \right) \right) &= 0,
\end{aligned}
\right\}
$$

uniformly in $X \times X$. Then there exists a unique multiplicative inverse vigintic map-
ping $\mathcal{M}_v : X \longrightarrow Y$ satisfying (5.2) such that

$$
\left.
\begin{aligned}
\lim_{t \to \infty} \mu \left(\mathcal{M}_v(x) - M_V(x), \frac{3 \cdot 3^{20} c_3 t \, \|x\|^\alpha}{2 \left(1 - 3^{\alpha+20} \right)} \right) &= 1 \\
\lim_{t \to \infty} \nu \left(\mathcal{M}_v(x) - M_V(x), \frac{3 \cdot 3^{20} c_3 t \, \|x\|^\alpha}{2 \left(1 - 3^{\alpha+20} \right)} \right) &= 0,
\end{aligned}
\right\}
$$

uniformly in X.

5.4 Counter-Examples

In this section, we show that the Eqs. (5.1) and (5.2) are not valid for $p = -19$
in Corollary 5.3.2 and $p = -20$ in Corollary 5.3.6, respectively, in the setting of
non-zero real numbers.

Example 5.4.1 Let us define the function

$$
\chi(x) =
\begin{cases}
\frac{c}{x^{19}}, & \text{for } x \in (1, \infty) \\
c, & \text{elsewhere}
\end{cases}
\tag{5.20}
$$

where $\chi : \mathbb{R}^* \longrightarrow \mathbb{R}$. Let $M_N : \mathbb{R}^* \longrightarrow \mathbb{R}$ be a function defined as

$$M_N(x) = \sum_{m=0}^{\infty} 1162261467^{-m} \chi(3^{-m}x) \tag{5.21}$$

for all $x \in \mathbb{R}$. Suppose the mapping $M_N : \mathbb{R}^* \longrightarrow \mathbb{R}$ described in (5.21) satisfies the inequality

$$|D_1 M_N(x, y)| \le \frac{1743392201 \, c}{581130733} \left(|x|^{-19} + |y|^{-19}\right) \tag{5.22}$$

for all $x, y \in \mathbb{R}^*$. We prove that there do not exist a multiplicative inverse novemdecic mapping $\mathcal{M}_n : \mathbb{R}^* \longrightarrow \mathbb{R}$ and a constant $\delta > 0$ such that

$$|M_N(x) - \mathcal{M}_n(x)| \le \delta \, |x|^{-19} \tag{5.23}$$

for all $x \in \mathbb{R}^*$. Firstly, let us prove that M_N satisfies (5.22). Using (5.20), we have

$$|M_N(x)| = \left| \sum_{m=0}^{\infty} 1162261467^{-m} \chi(3^{-m}x) \right| \le \sum_{m=0}^{\infty} \frac{c}{1162261467^m} = \frac{1162261467}{1162261466} c.$$

We find that that M_N is bounded by $\frac{1162261467 \, c}{1162261466}$ on $\mathbb{R}$. If $|x|^{-19} + |y|^{-19} \ge 1$, then the left hand side of (5.22) is less than $\frac{1743392201 \, c}{581130733}$. Now, suppose that $0 < |x|^{-19} + |y|^{-19} < 1$. Hence, there exists a positive integer m such that

$$\frac{1}{1162261467^{m+1}} \le |x|^{-19} + |y|^{-19} < \frac{1}{1162261467^m}. \tag{5.24}$$

Hence, the inequality (5.24) produces $1162261467^m \left(|x|^{-19} + |y|^{-19}\right) < 1$, or equivalently; $1162261467^m x^{-19} < 1$, $1162261467^m y^{-19} < 1$. So, $\frac{x^{19}}{1162261467^m} > 1$, $\frac{y^{19}}{1162261467^m} > 1$. Hence, the last inequalities imply $\frac{x^{19}}{1162261467^{m-1}} > 1162261467 > 1$, $\frac{y^{19}}{1162261467^{m-1}} > 1162261467 > 1$ and as a result, we find $\frac{1}{3^{m-1}}(x) > 1$, $\frac{1}{3^{m-1}}(y) > 1$, $\frac{1}{3^{m-1}}(2x + y) > 1$, $\frac{1}{3^{m-1}}(2x - y) > 1$.

Hence, for every value of $m = 0, 1, 2, \ldots, n - 1$, we obtain

$$\frac{1}{3^n}(x) > 1, \; \frac{1}{3^n}(y) > 1, \; \frac{1}{3^n}(2x + y) > 1, \; \frac{1}{3^n}(2x - y) > 1$$

and $D_1 M_N(3^{-n}x, 3^{-n}y) = 0$ for $m = 0, 1, 2, \ldots, n - 1$. Applying (5.20) and the definition of M_N, we obtain

$|D_1 M_N(x, y)|$

$$\leq \sum_{m=n}^{\infty} \frac{c}{1162261467^m} + \sum_{m=n}^{\infty} \frac{c}{1162261467^m} + \frac{1162261468}{1162261467} \sum_{m=n}^{\infty} \frac{c}{1162261467^m}$$

$$\leq \frac{3486784402\, c}{1162261467} \frac{1}{1162261467^m} \left(1 - \frac{1}{1162261467}\right)^{-1}$$

$$\leq \frac{3486784402\, c}{1162261466} \frac{1}{1162261467^{m+1}}$$

$$\leq \frac{1743392201\, c}{581130733} \left(|x|^{-19} + |y|^{-19}\right)$$

for all $x, y \in \mathbb{R}^*$. This means that the inequality (5.22) holds. We claim that the multiplicative inverse novemdecic functional equation (5.1) is unstable for $p = -19$ in Corollary 5.3.2. Assume that there exists a multiplicative inverse novemdecic mapping $M_N : \mathbb{R}^* \longrightarrow \mathbb{R}$ satisfying (5.23). So, we have

$$|M_N(x)| \leq (\delta + 1)|x|^{-19}. \tag{5.25}$$

Moreover, it is possible to choose a positive integer m with the condition $mc > \delta + 1$. If $x \in \left(1, 3^{m-1}\right)$, then $3^{-n}x \in (1, \infty)$ for all $m = 0, 1, 2, \ldots, n - 1$ and thus

$$|M_N(x)| = \sum_{m=0}^{\infty} \frac{\chi(3^{-m}x)}{1162261467^m} \geq \sum_{m=0}^{n-1} \frac{\frac{1162261467^m c}{x^{19}}}{1162261467^m} = \frac{mc}{x^{19}} > (\delta + 1)x^{-19}$$

which contradicts (5.25). Therefore, the multiplicative inverse novemdecic functional equation (5.1) is unstable for $p = -19$ in Corollary 5.3.2.

Similar to Example 5.4.1, the following example acts as a counter-example that the Eq. (5.2) is not stable for $p = -20$ in Corollary 5.3.6.

Example 5.4.2 Define the function $\xi : \mathbb{R}^* \longrightarrow \mathbb{R}$ via

$$\xi(x) = \begin{cases} \frac{\lambda}{x^{20}} & \text{for } u \in (1, \infty) \\ c, & \text{otherwise} \end{cases}. \tag{5.26}$$

Let $M_V : \mathbb{R}^* \longrightarrow \mathbb{R}$ be defined by

$$M_V(x) = \sum_{m=0}^{\infty} 3486784401^{-m} \xi(3^{-m}x)$$

for all $x \in \mathbb{R}$. Suppose the function M_V satisfies the inequality

$$|D_2 M_V(x, y)| \leq \frac{2615088301\, \lambda}{871696100} \left(|x|^{-20} + |y|^{-20}\right)$$

for all $x, y \in \mathbb{R}^*$. Then, there do not exist a multiplicative inverse vigintic mapping $\mathcal{M}_v : \mathbb{R}^* \longrightarrow \mathbb{R}$ and a constant $\eta > 0$ such that

$$|M_V(x) - \mathcal{M}_v(x)| \leq \eta |x|^{-20}$$

for all $x \in \mathbb{R}^*$.

Chapter 6
Solution to the Ulam Stability Problem of Multiplicative Inverse Type Unvigintic and Duovigintic Functional Equations in Paranormed Spaces

Abstract In this chapter, the generalized Hyers-Ulam stability of multiplicative inverse unvigintic and duovigintic functional equations in paranormed spaces using direct and fixed point methods are presented. Counter-examples to invalidate the stability results for critical cases are also discussed.

6.1 Introduction

For the first time, the fixed point theorem is employed to solve stability problem of functional equation in [59]. In most of the stability problems of functional equations, we find that an exact solution of a given equation is directly generated as a limit of a Cauchy sequence and this method of solving stability problem is called as direct method. The Hyers-Ulam stability problem of different form of functional equations is dealt in [96, 97] in the setting of paranormed spaces using direct and fixed point methods.

In this chapter, we deal with the multiplicative inverse unvigintic functional equation

$$
\begin{aligned}
&f(2x + y) + f(2x - y) \\
&= \frac{4f(x)f(y)}{\left(4f(y)^{2/21} - f(x)^{2/21}\right)^{21}} \left[\frac{1}{2} \sum_{k=0}^{10} \binom{21}{2k} [f(x)]^{2k/21} [f(y)]^{(21-2k)/21} \right]
\end{aligned} \tag{6.1}
$$

and a multiplicative inverse duovigintic functional equation

$$
\begin{aligned}
&f(2x + y) + f(2x - y) \\
&= \frac{4f(x)f(y)}{\left(4f(y)^{1/11} - f(x)^{1/11}\right)^{22}} \left[\sum_{k=0}^{11} \binom{22}{2k} [f(x)]^{k/11} [f(y)]^{(22-2k)/22} \right].
\end{aligned} \tag{6.2}
$$

© The Editor(s) (if applicable) and The Author(s), under exclusive license
to Springer Nature Switzerland AG 2020
B. V. Senthil Kumar and H. Dutta, *Multiplicative Inverse Functional Equations*,
Studies in Systems, Decision and Control 289, https://doi.org/10.1007/978-3-030-45355-8_6

We apply direct method and fixed point method to investigate the generalized Hyers-Ulam stability of the functional equations (6.1) and (6.2). We also present counter-examples to disprove the stability results for singular cases. For the notational convenience, let us define

$$\Delta_1 f(x, y) = f(2x + y) + f(2x - y)$$

$$- \frac{4f(x)f(y)}{\left(4f(y)^{2/21} - f(x)^{2/21}\right)^{21}} \left[\frac{1}{2} \sum_{k=0}^{10} \binom{21}{2k} [f(x)]^{2k/21} [f(y)]^{(21-2k)/21} \right]$$

and

$$\Delta_2 f(x, y) = f(2x + y) + f(2x - y)$$

$$- \frac{4f(x)f(y)}{\left(4f(y)^{1/11} - f(x)^{1/11}\right)^{22}} \left[\sum_{k=0}^{11} \binom{22}{2k} [f(x)]^{k/11} [f(y)]^{(22-2k)/22} \right].$$

6.2   Preliminaries

In this section, we evoke fundamental concepts pertinent to Fréchet spaces and basic results of fixed point theory.

The notion associated with statistical convergence for sequences of real numbers was introduced in [33, 140] independently and since then various generalizations and applications of this idea have been investigated by various mathematicians (see [35, 69, 86, 87, 131]). This notion was defined in normed spaces by Kolk [70].

Definition 6.2.1 ([143]) Let X be a vector space. A *paranorm* $P : X \to [0, \infty)$ is a function on X such that

(1) $P(0) = 0$;
(2) $P(-x) = P(x)$;
(3) $P(x + y) \leq P(x) + P(y)$ (triangle inequality);
(4) If $\{t_n\}$ is a sequence of scalars with $t_n \to t$ and $\{x_n\} \subset X$ with $P(x_n\text{-}x) \to 0$, then $P(t_n x_n\text{-}tx) \to 0$ (continuity of multiplication);
     The pair (X, P) is called a *paranormed space* if P is a paranorm on X.
(5) $P(x) = 0$ implies $x = 0$.

A Fréchet space is a total and complete paranormed space.

Definition 6.2.2 Let A be a set. A function $d : A \times A \to [0, \infty]$ is called a *generalized metric* on A if d satisfies the following conditions:

1. $d(x, y) = 0$ if and only if $x = y$;
2. $d(x, y) = d(y, x)$ for all $x, y \in A$;
3. $d(x, z) \leq d(x, y) + d(y, z)$ for all $x, y, z \in A$.

We note that the only one difference of the generalized metric from the usual metric is that the range of the former is permitted to include infinity. The ensuing theorem is very useful for proving our main results which is presented in [75].

Theorem 6.2.3 ([75]) *Suppose (X, d) is a complete generalized metric space and let a mapping $J : X \longrightarrow X$ be strictly contractive with Lipschitz constant $L < 1$. Then for each given element $X \in X$, either*

$$d\left(J^n x, J^{n+1} x\right) = \infty$$

for all non-negative integers n or there exists a positive integer n_0 such that

1. *$d\left(J^n x, J^{n+1} x\right) < \infty$ for all $n \geq n_0$;*
2. *the sequence $\{J^n x\}$ converges to a fixed point y^* of J;*
3. *y^* is the unique fixed point of J in the set $Y = \{y \in X \mid d\left(J^{n_0} x, y\right) < \infty\}$;*
4. *$d\left(y, y^*\right) < \frac{1}{1-L} d(y, Jy)$, for all $y \in Y$.*

In this entire chapter, let us assume that (X, P) as a Fréchet space and $(Y, \|\cdot\|)$ as a Banach space.

6.3 Stability of (6.1): Direct Method

In this section, we investigate the generalized Hyers-Ulam stability of (6.1) and then we extend its T. M. Rassias stability, J.M. Rassias stabilities in the consequent corollaries.

Theorem 6.3.1 *Let $\phi : Y \times Y \longrightarrow [0, \infty)$ be a function satisfying*

$$\sum_{i=0}^{\infty} 3^{21i} \phi\left(3^i x, 3^i y\right) < +\infty \tag{6.3}$$

for all $x, y \in Y$. If a function $f : Y \longrightarrow X$ satisfies the functional inequality

$$P\left(\Delta_1 f(x, y)\right) \leq \phi(x, y) \tag{6.4}$$

for all $x, y \in Y$, then there exists a unique multiplicative inverse unvigintic mapping $F : Y \longrightarrow X$ which satisfies (6.1) and the inequality

$$P(f(x) - F(x)) \leq \sum_{i=0}^{\infty} 3^{21i} \phi\left(3^i x, 3^i x\right) \tag{6.5}$$

for all $x \in Y$.

Proof First, considering (x, y) as (x, x) in (6.4) and on further simplification, we get

$$P(f(x) - 3^{21} f(3x)) \leq \phi(x, x) \tag{6.6}$$

for all $x \in Y$. Now, replacing x by $3x$ in (6.6), multiplying by 3^{21} and summing the resulting inequality with (6.6), we obtain

$$P\left(f(x) - 3^{42} f\left(3^2 x\right)\right) \leq \sum_{i=0}^{1} 3^{21i} \phi\left(3^i x, 3^i x\right)$$

for all $x \in Y$. Proceeding further and using induction arguments on a positive integer n, we arrive

$$P\left(f(x) - 3^{21n} f\left(3^n x\right)\right) \leq \sum_{i=0}^{n-1} 3^{21i} \phi\left(3^i x, 3^i x\right) \tag{6.7}$$

for all $x \in Y$. Hence for any non-negative integers l, k with $l > k$, we obtain by using the triangle inequality

$$P\left(3^{21l} f\left(3^l x\right) - 3^{21k} f\left(3^k x\right)\right) \leq P\left(3^{21l} f\left(3^l x\right) - f(x)\right) + P\left(f(x) - 3^{21k} f\left(3^k x\right)\right)$$

$$\leq \sum_{i=0}^{l-1} 3^{21i} \phi\left(3^i x, 3^i x\right) + \sum_{i=0}^{k-1} 3^{21i} \phi\left(3^i x, 3^i x\right)$$

$$\leq \sum_{i=k}^{l-1} 3^{21i} \phi\left(3^i x, 3^i x\right) \tag{6.8}$$

for all $x \in Y$. Taking the limit as $k \to +\infty$ in (6.8) and considering (6.3), it follows that the sequence $f_n(x) = \{3^{21n} f(3^n x)\}$ is a Cauchy sequence for each $x \in Y$. Since X is complete, we can define $F : Y \longrightarrow X$ by

$$F(x) = \lim_{n \to \infty} 3^{21n} f\left(3^n x\right). \tag{6.9}$$

To show that F satisfies (6.1), replacing (x, y) by $(3^n x, 3^n y)$ in (6.4) and multiplying by 3^{21n}, we obtain

$$P\left(3^{21n} \Delta_1 f\left(3^n x, 3^n y\right)\right) \leq 3^{21n} \phi\left(3^n x, 3^n y\right) \tag{6.10}$$

for all $x, y \in Y$, for all positive integer n. Using (6.3) and (6.9) in (6.10), we see that F satisfies (6.1), for all $x, y \in Y$. Taking limit $n \to \infty$ in (6.7), we arrive (6.5). Now, it remains to show that F is uniquely defined. Let $F' : Y \longrightarrow X$ be another multiplicative inverse unvigintic mapping which satisfies (6.1) and the inequality (6.5). Then we have

$$P(F(x) - f'(x)) = P\left(3^{21n} F\left(3^n x\right) - 3^{21n} f'\left(3^n x\right)\right)$$
$$\leq P\left(3^{21n} F\left(3^n x\right) - 3^{21n} F'\left(3^n x\right)\right) + P\left(3^{21n} F\left(3^n x\right) - 3^{21n} F'\left(3^n x\right)\right)$$
$$\leq 2 \sum_{i=0}^{\infty} 3^{21(n+i)} \phi\left(3^{n+i} x, 3^{n+i} x\right)$$
$$\leq 2 \sum_{i=n}^{\infty} 3^{21i} \phi\left(3^i x, 3^i x\right) \tag{6.11}$$

for all $x \in Y$. Allowing $n \to \infty$ in (6.11), we see that F is unique, which completes the proof of Theorem 6.3.1. $\qquad\square$

Theorem 6.3.2 *Let $\phi : Y \times Y \longrightarrow [0, \infty)$ be a function satisfying*

$$\sum_{i=0}^{\infty} \frac{1}{3^{21(i+1)}} \phi\left(\frac{x}{3^{i+1}}, \frac{y}{3^{i+1}}\right) < +\infty$$

for all $x, y \in Y$. If a function $f : Y \longrightarrow X$ satisfies the functional inequality (6.4), for all $x, y \in Y$, then there exists a unique multiplicative inverse unvigintic mapping $F : Y \longrightarrow X$ which satisfies (6.1) and the inequality

$$P(f(x) - F(x)) \leq \sum_{i=0}^{\infty} \frac{1}{3^{21(i+1)}} \phi\left(\frac{x}{3^{i+1}}, \frac{x}{3^{i+1}}\right)$$

for all $x \in Y$.

Proof The proof is obtained by replacing (x, y) by $\left(\frac{x}{3}, \frac{x}{3}\right)$ in (6.4) and proceeding by similar arguments as in Theorem 6.3.1. $\qquad\square$

The following corollaries are immediate consequences of Theorems (6.3.1) and (6.3.2). Hence we omit the proof.

Corollary 6.3.3 *Let $f : Y \longrightarrow X$ be a mapping and let there exist real numbers $\alpha \neq -21$ and $c_1 \geq 0$ such that*

$$P\left(\Delta_1 f(x, y)\right) \leq c_1 \left(\|x\|^{\alpha} + \|y\|^{\alpha}\right) \tag{6.12}$$

for all $x, y \in Y$. Then there exists a unique multiplicative inverse unvigintic mapping $F : Y \to X$ satisfying (6.1) and

$$P(f(x) - F(x)) \leq \begin{cases} \frac{9c_1}{1 - 3^{\alpha+21}} \|x\|^{\alpha} & \text{for } \alpha < -21 \\ \frac{9c_1}{3^{\alpha+21} - 1} \|x\|^{\alpha} & \text{for } \alpha > -21 \end{cases} \tag{6.13}$$

for every $x \in Y$.

Corollary 6.3.4 *Let $f : Y \longrightarrow X$ be a mapping and let there exist real numbers a, b such that $\rho = a + b \neq -21$. Let there exist $c_2 \geq 0$ such that*

$$P\left(\Delta_1 f(x, y)\right) \leq c_2 \|x\|^a \|y\|^b \tag{6.14}$$

for all $x, y \in Y$. Then there exists a unique multiplicative inverse unvigintic mapping $F : Y \longrightarrow X$ satisfying (6.1) and

$$P(f(x) - F(x)) \leq \begin{cases} \frac{3c_2}{1-3^{\rho+21}} \|x\|^\rho & \text{for } \rho < -21 \\ \frac{3c_2}{3^{\rho+21}-1} \|x\|^\rho & \text{for } \rho > -21 \end{cases} \tag{6.15}$$

for every $x \in Y$.

Corollary 6.3.5 *Let $c_3 \geq 0$ and p, q be real numbers such that $\lambda = p + q \neq -21$, and $f : Y \longrightarrow X$ be a mapping satisfying the functional inequality*

$$P\left(\Delta_1 f(x, y)\right) \leq c_3 \left(\|x\|^p \|y\|^q + + \left(\|x\|^{p+q} + \|y\|^{p+q}\right)\right) \tag{6.16}$$

for all $x, y \in Y$. Then there exists a unique multiplicative inverse unvigintic mapping $F : Y \longrightarrow X$ satisfying (6.1) and

$$P(f(x) - F(x)) \leq \begin{cases} \frac{9c_3}{1-3^{\lambda+21}} \|x\|^\lambda & \text{for } \lambda < -21 \\ \frac{9c_3}{3^{\lambda+21}-1} \|x\|^\lambda & \text{for } \lambda > -21 \end{cases} \tag{6.17}$$

for every $x \in Y$.

Theorem 6.3.6 *Let $\phi : X \times X \longrightarrow [0, \infty)$ be a function satisfying*

$$\sum_{i=0}^{\infty} 3^{21i} \phi\left(3^i x, 3^i y\right) < +\infty$$

for all $x, y \in X$. If a function $f : X \longrightarrow Y$ satisfies the functional inequality

$$\|\Delta_1 f(x, y)\| \leq \phi(x, y) \tag{6.18}$$

for all $x, y \in X$, then there exists a unique multiplicative inverse unvigintic mapping $F : X \longrightarrow Y$ which satisfies (6.1) and the inequality

$$\|f(x) - F(x)\| \leq \sum_{i=0}^{\infty} 3^{21i} \phi\left(3^i x, 3^i x\right)$$

for all $x \in X$.

Proof The proof is obtained by similar arguments as in Theorem 6.3.1. $\qquad\square$

Theorem 6.3.7 *Let* $\phi : X \times X \longrightarrow [0, \infty)$ *be a function satisfying*

$$\sum_{i=0}^{\infty} \frac{1}{3^{21(i+1)}} \phi \left(\frac{x}{3^{i+1}}, \frac{y}{2^{i+1}} \right) < +\infty$$

for all $x, y \in X$. *If a function* $f : X \longrightarrow Y$ *satisfies the functional inequality (6.18) for all* $x, y \in X$, *then there exists a unique multiplicative inverse unvigintic mapping* $F : X \longrightarrow Y$ *which satisfies (6.1) and the inequality*

$$\| f(x) - F(x) \| \leq \sum_{i=0}^{\infty} \frac{1}{3^{21(i+1)}} \phi \left(\frac{x}{3^{i+1}}, \frac{x}{3^{i+1}} \right)$$

for all $x \in X$.

Proof The proof is analogous to the proof of Theorem 6.3.2.   □

Corollary 6.3.8 *Let* $f : X \longrightarrow Y$ *be a mapping and let there exist real numbers* $\alpha < -21$ *and* $c_1 \geq 0$ *such that*

$$\| \Delta_1 f(x, y) \| \leq c_1 \left(P(x)^\alpha + P(y)^\alpha \right) \tag{6.19}$$

for all $x, y \in X$. *Then there exists a unique multiplicative inverse unvigintic mapping* $F : X \longrightarrow Y$ *satisfying (6.1) and*

$$\| f(x) - F(x) \| \leq \frac{9 c_1}{1 - 3^{\alpha + 21}} P(x)^\alpha \tag{6.20}$$

for every $x \in X$.

Proof The proof follows immediately by taking $\phi(x, y) = c_1 \left(P(x)^\alpha + P(y)^\alpha \right)$, for all $x, y \in X$ in Theorem 6.3.6.   □

Corollary 6.3.9 *Let* $f : X \longrightarrow Y$ *be a mapping and let there exist real numbers* a, b *such that* $\rho = a + b < -21$. *Let there exist* $c_2 \geq 0$ *such that*

$$\| \Delta_1 f(x, y) \| \leq c_2 P(x)^a P(y)^b \tag{6.21}$$

for all $x, y \in X$. *Then there exists a unique multiplicative inverse unvigintic mapping* $F : X \longrightarrow Y$ *satisfying (6.1) and*

$$\| f(x) - F(x) \| \leq \frac{3 c_2}{1 - 3^{\rho + 21}} P(x)^\rho \tag{6.22}$$

for every $x \in X$.

Proof The required results in Corollary 6.3.9 can be easily derived by considering $\phi(x, y) = c_2 P(x)^a P(y)^b$, for all $x, y \in X$ in Theorem 6.3.6.   □

Corollary 6.3.10 *Let $c_3 \geq 0$ and p, q be real numbers such that $\lambda = p + q < -21$, and $f : X \longrightarrow Y$ be a mapping satisfying the functional inequality*

$$\|\Delta_1 f(x, y)\| \leq c_3 \left(P(x)^p P(y)^q + \left(P(x)^{p+q} + P(y)^{p+q} \right) \right) \tag{6.23}$$

for all $x, y \in X$. Then there exists a unique multiplicative inverse unvigintic mapping $F : X \longrightarrow Y$ satisfying (6.1) and

$$\|f(x) - F(x)\| \leq \frac{9c_3}{1 - 3^{\lambda+21}} P(x)^\lambda \tag{6.24}$$

for every $x \in X$.

Proof By choosing $\phi(x, y) = c_3 \left(P(x)^p P(y)^q + \left(P(x)^{p+q} + P(y)^{p+q} \right) \right)$, for all $x, y \in X$ in Theorem 6.3.6, the proof of Corollary 6.3.10 is complete. $\qquad\square$

6.4 Stability of (6.2): Direct Method

In this section, we investigate the generalized Hyers-Ulam stability of (6.2) and then we extend its T. M. Rassias stability, J.M. Rassias stabilities in the consequent corollaries.

Theorem 6.4.1 *Let $\phi : Y \times Y \longrightarrow [0, \infty)$ be a function satisfying*

$$\sum_{i=0}^{\infty} 3^{22i} \phi \left(3^i x, 3^i y \right) < +\infty \tag{6.25}$$

for all $x, y \in Y$. If a function $f : Y \longrightarrow X$ satisfies the functional inequality

$$P \left(\Delta_2 f(x, y) \right) \leq \phi(x, y) \tag{6.26}$$

for all $x, y \in Y$, then there exists a unique multiplicative inverse duovigintic mapping $F : Y \longrightarrow X$ which satisfies (6.2) and the inequality

$$P(f(x) - F(x)) \leq \sum_{i=0}^{\infty} 3^{22i} \phi \left(3^i x, 3^i x \right) \tag{6.27}$$

for all $x \in Y$.

Proof First, considering (x, y) as (x, x) in (6.26) and on further simplification, we get

$$P(f(x) - 3^{22} f(3x)) \leq \phi(x, x) \tag{6.28}$$

for all $x \in Y$. Now, replacing x by $3x$ in (6.28), multiplying by 3^{22} and summing the resulting inequality with (6.28), we obtain

$$P\left(f(x) - 3^{44} f\left(3^2 x\right)\right) \leq \sum_{i=0}^{1} 3^{22i} \phi\left(3^i x, 3^i x\right)$$

for all $x \in Y$. Proceeding further and using induction arguments on a positive integer n, we arrive

$$P\left(f(x) - 3^{22n} f\left(3^n x\right)\right) \leq \sum_{i=0}^{n-1} 3^{22i} \phi\left(3^i x, 3^i x\right) \tag{6.29}$$

for all $x \in Y$. Hence for any non-negative integers l, k with $l > k$, we obtain by using the triangle inequality

$$P\left(3^{22l} f\left(3^l x\right) - 3^{22k} f\left(3^k x\right)\right) \leq P\left(3^{22l} f\left(3^l x\right) - f(x)\right) + P\left(f(x) - 3^{22k} f\left(3^k x\right)\right)$$

$$\leq \sum_{i=0}^{l-1} 3^{22i} \phi\left(3^i x, 3^i x\right) + \sum_{i=0}^{k-1} 3^{22i} \phi\left(3^i x, 3^i x\right)$$

$$\leq \sum_{i=k}^{l-1} 3^{22i} \phi\left(3^i x, 3^i x\right) \tag{6.30}$$

for all $x \in Y$. Taking the limit as $k \to +\infty$ in (6.30) and considering (6.25), it follows that the sequence $f_n(x) = \{3^{22n} f(3^n x)\}$ is a Cauchy sequence for each $x \in Y$. Since X is complete, we can define $F : Y \longrightarrow X$ by

$$F(x) = \lim_{n \to \infty} 3^{22n} f\left(3^n x\right). \tag{6.31}$$

To show that F satisfies (6.2), replacing (x, y) by $(3^n x, 3^n y)$ in (6.26) and multiplying by 3^{22n}, we obtain

$$P\left(3^{22n} \Delta_2 f\left(3^n x, 3^n y\right)\right) \leq 3^{22n} \phi\left(3^n x, 3^n y\right) \tag{6.32}$$

for all $x, y \in Y$, for all positive integer n. Using (6.25) and (6.29) in (6.31), we see that F satisfies (6.2), for all $x, y \in Y$. Taking limit $n \to \infty$ in (6.29), we arrive (6.27). Now, it remains to show that F is uniquely defined. Let $F' : Y \longrightarrow X$ be another multiplicative inverse duovigintic mapping which satisfies (6.2) and the inequality (6.27). Then we have

$$P(F(x) - F'(x)) = P\left(3^{22n} F\left(3^n x\right) - 3^{22n} F'\left(3^n x\right)\right)$$
$$\leq P\left(3^{22n} F\left(3^n x\right) - 3^{22n} F'\left(3^n x\right)\right) + P\left(3^{22n} F\left(3^n x\right) - 3^{22n} F'\left(3^n x\right)\right)$$
$$\leq 2\sum_{i=0}^{\infty} 3^{22(n+i)} \phi\left(3^{n+i} x, 3^{n+i} x\right)$$
$$\leq 2\sum_{i=n}^{\infty} 3^{22i} \phi\left(3^i x, 3^i x\right) \tag{6.33}$$

for all $x \in Y$. Allowing $n \to \infty$ in (6.33), we see that F is unique, which completes the proof of Theorem 6.4.1. $\qquad\square$

Theorem 6.4.2 *Let $\phi : Y \times Y \longrightarrow [0, \infty)$ be a function satifying*

$$\sum_{i=0}^{\infty} \frac{1}{3^{22(i+1)}} \phi\left(\frac{x}{3^{i+1}}, \frac{y}{3^{i+1}}\right) < +\infty$$

for all $x, y \in Y$. If a function $f : Y \longrightarrow X$ satisfies the functional inequality (6.26), for all $x, y \in Y$, then there exists a unique multiplicative inverse duovigintic mapping $F : Y \longrightarrow X$ which satisfies (6.2) and the inequality

$$P(f(x) - F(x)) \leq \sum_{i=0}^{\infty} \frac{1}{3^{22(i+1)}} \phi\left(\frac{x}{3^{i+1}}, \frac{x}{3^{i+1}}\right)$$

for all $x \in Y$.

Proof The proof is obtained by replacing (x, y) by $\left(\frac{x}{3}, \frac{x}{3}\right)$ in (6.26) and proceeding by similar arguments as in Theorem 6.4.1. $\qquad\square$

The following corollaries are immediate consequences of Theorems (6.4.1) and (6.4.2). Hence we omit the proof.

Corollary 6.4.3 *Let $f : Y \longrightarrow X$ be a mapping and let there exist real numbers $\alpha \neq -22$ and $c_1 \geq 0$ such that*

$$P\left(\Delta_2 f(x, y)\right) \leq c_1\left(\|x\|^\alpha + \|y\|^\alpha\right) \tag{6.34}$$

for all $x, y \in Y$. Then there exists a unique multiplicative inverse duovigintic mapping $F : Y \to X$ satisfying (6.2) and

$$P(f(x) - F(x)) \leq \begin{cases} \frac{9c_1}{1 - 3^{\alpha+22}} \|x\|^\alpha & \text{for } \alpha < -22 \\ \frac{9c_1}{3^{\alpha+22} - 1} \|x\|^\alpha & \text{for } \alpha > -22 \end{cases} \tag{6.35}$$

for every $x \in Y$.

Corollary 6.4.4 *Let* $f : Y \longrightarrow X$ *be a mapping and let there exist real numbers* a, b *such that* $\rho = a + b \neq -22$. *Let there exist* $c_2 \geq 0$ *such that*

$$P(\Delta_2 f(x, y)) \leq c_2 \|x\|^a \|y\|^b \tag{6.36}$$

for all $x, y \in Y$. *Then there exists a unique multiplicative inverse duovigintic mapping* $F : Y \longrightarrow X$ *satisfying (6.2) and*

$$P(f(x) - F(x)) \leq \begin{cases} \frac{3c_2}{1 - 3^{\rho + 22}} \|x\|^\rho & \text{for } \rho < -22 \\ \frac{3c_2}{3^{\rho + 22} - 1} \|x\|^\rho & \text{for } \rho > -22 \end{cases} \tag{6.37}$$

for every $x \in Y$.

Corollary 6.4.5 *Let* $c_3 \geq 0$ *and* p, q *be real numbers such that* $\lambda = p + q \neq -22$, *and* $f : Y \longrightarrow X$ *be a mapping satisfying the functional inequality*

$$P(\Delta_2 f(x, y)) \leq c_3 \left(\|x\|^p \|y\|^q + + \left(\|x\|^{p+q} + \|y\|^{p+q} \right) \right) \tag{6.38}$$

for all $x, y \in Y$. *Then there exists a unique multiplicative inverse duovigintic mapping* $F : Y \longrightarrow X$ *satisfying (6.2) and*

$$P(f(x) - F(x)) \leq \begin{cases} \frac{9c_3}{1 - 3^{\lambda + 22}} \|x\|^\lambda & \text{for } \lambda < -22 \\ \frac{9c_3}{3^{\lambda + 22} - 1} \|x\|^\lambda & \text{for } \lambda > -22 \end{cases} \tag{6.39}$$

for every $x \in Y$.

Theorem 6.4.6 *Let* $\phi : X \times X \longrightarrow [0, \infty)$ *be a function satisfying*

$$\sum_{i=0}^{\infty} 3^{22i} \phi \left(3^i x, 3^i y \right) < +\infty$$

for all $x, y \in X$. *If a function* $f : X \longrightarrow Y$ *satisfies the functional inequality*

$$\|\Delta_2 f(x, y)\| \leq \phi(x, y) \tag{6.40}$$

for all $x, y \in X$, *then there exists a unique multiplicative inverse duovigintic mapping* $F : X \longrightarrow Y$ *which satisfies (6.2) and the inequality*

$$\|f(x) - F(x)\| \leq \sum_{i=0}^{\infty} 3^{22i} \phi \left(3^i x, 3^i x \right)$$

for all $x \in X$.

Proof The proof is obtained by similar arguments as in Theorem 6.4.1. $\qquad\square$

Theorem 6.4.7 *Let $\phi : X \times X \longrightarrow [0, \infty)$ be a function satisfying*

$$\sum_{i=0}^{\infty} \frac{1}{3^{22(i+1)}} \phi\left(\frac{x}{3^{i+1}}, \frac{y}{2^{i+1}}\right) < +\infty$$

for all $x, y \in X$. If a function $f : X \longrightarrow Y$ satisfies the functional inequality (6.40) for all $x, y \in X$, then there exists a unique multiplicative inverse duovigintic mapping $F : X \longrightarrow Y$ which satisfies (6.2) and the inequality

$$\|f(x) - F(x)\| \leq \sum_{i=0}^{\infty} \frac{1}{3^{22(i+1)}} \phi\left(\frac{x}{3^{i+1}}, \frac{x}{3^{i+1}}\right)$$

for all $x \in X$.

Proof The proof is analogous to the proof of Theorem 6.4.2.   □

Corollary 6.4.8 *Let $f : X \longrightarrow Y$ be a mapping and let there exist real numbers $\alpha < -22$ and $c_1 \geq 0$ such that*

$$\|\Delta_2 f(x, y)\| \leq c_1 \left(P(x)^\alpha + P(y)^\alpha\right) \tag{6.41}$$

for all $x, y \in X$. Then there exists a unique multiplicative inverse duovigintic mapping $F : X \longrightarrow Y$ satisfying (6.2) and

$$\|f(x) - F(x)\| \leq \frac{9c_1}{1 - 3^{\alpha+22}} P(x)^\alpha \tag{6.42}$$

for every $x \in X$.

Proof The proof follows immediately by taking $\phi(x, y) = c_1 \left(P(x)^\alpha + P(y)^\alpha\right)$, for all $x, y \in X$ in Theorem 6.4.6.   □

Corollary 6.4.9 *Let $f : X \longrightarrow Y$ be a mapping and let there exist real numbers a, b such that $\rho = a + b < -22$. Let there exist $c_2 \geq 0$ such that*

$$\|\Delta_2 f(x, y)\| \leq c_2 P(x)^a P(y)^b \tag{6.43}$$

for all $x, y \in X$. Then there exists a unique multiplicative inverse duovigintic mapping $F : X \longrightarrow Y$ satisfying (6.2) and

$$\|f(x) - F(x)\| \leq \frac{3c_2}{1 - 3^{\rho+22}} P(x)^\rho \tag{6.44}$$

for every $x \in X$.

Proof The required results in Corollary 6.4.9 can be easily derived by considering $\phi(x, y) = c_2 P(x)^a P(y)^b$, for all $x, y \in X$ in Theorem 6.4.6.   □

Corollary 6.4.10 *Let $c_3 \geq 0$ and p, q be real numbers such that $\lambda = p + q < -22$, and $f : X \longrightarrow Y$ be a mapping satisfying the functional inequality*

$$\|\Delta_2 f(x, y)\| \leq c_3 \left(P(x)^p P(y)^q + \left(P(x)^{p+q} + P(y)^{p+q}\right)\right) \tag{6.45}$$

for all $x, y \in X$. Then there exists a unique multiplicative inverse duovigintic mapping $F : X \longrightarrow Y$ satisfying (6.2) and

$$\|f(x) - F(x)\| \leq \frac{9c_3}{1 - 3^{\lambda+22}} P(x)^\lambda \tag{6.46}$$

for every $x \in X$.

Proof By choosing $\phi(x, y) = c_3 \left(P(x)^p P(y)^q + \left(P(x)^{p+q} + P(y)^{p+q}\right)\right)$, for all $x, y \in X$ in Theorem 6.4.6, the proof of Corollary 6.4.10 is complete. □

6.5   Stability of (6.1): Fixed Point Method

Theorem 6.5.1 *Suppose that the mapping $f : Y \to X$ satisfies the inequality*

$$P(\Delta_1 f(x, y)) \leq \psi(x, y) \tag{6.47}$$

for all $x, y \in Y$, where $\psi : Y \times Y \longrightarrow [0, \infty)$ is a given function. If there exists $L < 1$ such that

$$\psi(x, y) \leq \frac{1}{3^{21}} L \psi \left(\frac{x}{3}, \frac{y}{3}\right) \tag{6.48}$$

for all $x, y \in Y$, then there exists a unique multiplicative inverse unvigintic mapping $F : Y \longrightarrow X$ such that

$$P(F(x) - f(x)) \leq \frac{L}{1 - L} \psi \left(\frac{x}{3}, \frac{x}{3}\right) \tag{6.49}$$

for all $x \in Y$.

Proof Define a set S by

$$S = \{h : Y \longrightarrow X | h \text{ is a function}\}$$

and introduce the generalized metric d on S as follows:

$$d(g, h) = \inf\{C \in \mathbb{R}_+ : P(g(x) - h(x)) \leq C\psi(x, x), \text{ for all } x \in Y\} \tag{6.50}$$

where, as usual, $\inf \phi = +\infty$. It is easy to show that (S, d) is complete (see Lemma 2.1).

Define a mapping $\sigma : S \longrightarrow S$ by

$$\sigma h(x) = 3^{21} h(3x) \quad (x \in Y) \tag{6.51}$$

for all $h \in S$. We claim that σ is strictly contractive on S. For any given $g, h \in S$, let $C_{gh} \in [0, \infty]$ be an arbitrary constant with $d(g, h) \le C_{gh}$. Hence

$$\begin{aligned}
d(g, h) < C_{gh} &\Rightarrow P(g(x) - h(x)) \le C_{gh} \psi(x, x), \forall x \in Y \\
&\Rightarrow P(3^{21} g(3x) - 3^{21} h(3x)) \le 3^{21} C_{gh} \psi(3x, 3x), \forall x \in Y \\
&\Rightarrow P(3^{21} g(3x) - 3^{21} h(3x)) \le L C_{gh} \psi(x, x), \forall x \in Y \\
&\Rightarrow d(\sigma g, \sigma h) \le L C_{gh}.
\end{aligned}$$

Therefore, we see that

$$d(\sigma g, \sigma h) \le L d(g, h), \text{ for all } g, h \in S$$

that is, σ is strictly contractive mapping of S, with the Lipschitz constant L.

Now, replacing (x, y) by (x, x) in (6.47) and simplifying further, we get

$$P(f(x) - 3^{21} f(3x)) \le 3^{21} \psi(x, x) \le L \psi \left(\frac{x}{3}, \frac{x}{3} \right)$$

for all $x \in Y$. Hence (6.50) implies that $d(f, \sigma f) \le 1$. Hence by applying the fixed point alternative Theorem 6.2.3, there exists a function $F : Y \longrightarrow X$ satisfying the following:

(1) F is a fixed point of σ, that is

$$F(3x) = \frac{1}{3^{21}} F(x) \tag{6.52}$$

for all $x \in Y$. The mapping F is the unique fixed point of σ in the set

$$\mu = \{g \in S : d(f, g) < \infty\}.$$

This implies that r is the unique mapping satisfying (6.52) such that there exists $C \in (0, \infty)$ satisfying

$$P(f(x) - F(x)) \le C \psi(x, x), \forall x \in Y.$$

(2) $d\left(\sigma^n f, r\right) \to 0$ as $n \to \infty$. Thus we have

$$\lim_{n\to\infty} 3^{21n} f\left(3^n x\right) = F(x) \tag{6.53}$$

for all $x \in Y$.

(3) $d(r, f) \le \frac{L}{1-L} d(r, \sigma f)$ which implies

$$d(r, f) \le \frac{L}{1 - L}.$$

Thus the inequality (6.49) holds. Hence from (6.47), (6.48) and (6.53), we have

$$
\begin{aligned}
P\left(\Delta_1 F(x, y)\right) &= \lim_{n\to\infty} 3^{21n} P\left(\Delta_1 f\left(3^n x, 3^n y\right)\right) \\
&\le \lim_{n\to\infty} 3^{21n} \psi\left(3^n x, 3^n y\right) \\
&\le \lim_{n\to\infty} 3^{21n} \frac{L^n}{3^{21n}} \psi(x, y) = 0
\end{aligned}
$$

for all $x, y \in Y$. So $\Delta_1 F(x, y) = 0$, for all $x, y \in Y$. Hence F is a solution of equation (6.1). Hence, $F : Y \longrightarrow X$ is a multiplicative inverse unvigintic mapping.

Next, we show that F is the unique multiplicative inverse unvigintic mapping satisfying (6.1) and (6.49). Suppose, let $F' : Y \longrightarrow X$ be another multiplicative inverse unvigintic mapping satisfying (6.1) and (6.49). Then from (6.1), we have that F' is a fixed point of σ. Since $d(F, F') < \infty$, we have

$$F' \in S^* = \{g \in S \mid d(F, g) < \infty\}.$$

From Theorem 6.2.3(3) and since both F and F' are fixed points of σ, we have $F = F'$. Therefore F is unique. Hence, there exists a unique multiplicative inverse unvigintic mapping $F : Y \longrightarrow X$ satisfying (6.1) and (6.49), which completes the proof of Theorem 6.5.1. □

Theorem 6.5.2 *Suppose that the mapping $f : Y \longrightarrow X$ satisfies the inequality (6.47), for all $x, y \in Y$, where $\psi : Y \times Y \longrightarrow [0, \infty)$ is a given function. If there exists $L < 1$ such that*

$$\psi\left(\frac{x}{3}, \frac{y}{3}\right) \le 3^{21} L \psi(x, y) \tag{6.54}$$

for all $x, y \in Y$. Then there exists a unique multiplicative inverse unvigintic mapping $F : Y \longrightarrow X$ such that

$$P(f(x) - F(x)) \le \frac{1}{1 - L} \psi\left(\frac{x}{3}, \frac{x}{3}\right)$$

for all $x \in Y$.

Proof The proof is similar to the proof of Theorem 6.5.1.   □

Corollary 6.5.3 *Let* $f : Y \longrightarrow X$ *be a mapping and let there exist real numbers* $\alpha \neq -21$ *and* $c_1 \geq 0$ *such that (6.12) holds for all* $x, y \in Y$. *Then there exists a unique multiplicative inverse unvigintic mapping* $F : Y \longrightarrow X$ *satisfying (6.1) and (6.13), for every* $x \in Y$.

Proof The proof follows immediately by taking $\psi(x, y) = c_1 (\|x\|^{\alpha} + \|y\|^{\alpha})$, for all $x, y \in Y$ and $L = 3^{\alpha+21}$, $L = 3^{-\alpha-21}$ in Theorems 6.5.1 and 6.5.2 respectively.   □

Corollary 6.5.4 *Let* $f : Y \longrightarrow X$ *be a mapping and let there exist real numbers* a, b *such that* $\rho = a + b \neq -21$. *Let there exist* $c_2 \geq 0$ *such that (6.14) holds for all* $x, y \in Y$. *Then there exists a unique multiplicative inverse unvigintic mapping* $F : Y \longrightarrow X$ *satisfying (6.1) and (6.15), for every* $x \in Y$.

Proof The required results in Corollary 6.5.4 can be easily derived by considering $\psi(x, y) = c_2 \|x\|^a \|y\|^b$, for all $x, y \in Y$ and $L = 3^{\rho+21}$, $L = 3^{-\rho-21}$ in Theorems 6.5.1 and 6.5.2 respectively.   □

Corollary 6.5.5 *Let* $c_3 \geq 0$ *and* p, q *be real numbers such that* $\lambda = p + q \neq -21$, *and* $f : Y \longrightarrow X$ *be a mapping satisfying the functional inequality (6.16), for all* $x, y \in Y$. *Then there exists a unique multiplicative inverse unvigintic mapping* $F : Y \longrightarrow X$ *satisfying (6.1) and (6.17), for every* $x \in Y$.

Proof By choosing $\psi(x, y) = c_3 \left(\|x\|^p \|y\|^q + + (\|x\|^{p+q} + \|y\|^{p+q}) \right)$, for all $x, y \in Y$ and $L = 3^{\lambda+21}$, $L = 3^{-\lambda-21}$ in Theorems 6.5.1 and 6.5.2 respectively, the proof of Corollary 6.5.5 is complete.   □

Theorem 6.5.6 *Suppose that the mapping* $f : X \longrightarrow Y$ *satisfies the inequality*

$$\|\Delta_1 f(x, y)\| \leq \psi(x, y) \tag{6.55}$$

for all $x, y \in X$, *where* $\psi : X \times X \longrightarrow [0, \infty)$ *is a given function. If there exists* $L < 1$ *such that (6.48) holds for all* $x, y \in X$, *then there exists a unique multiplicative inverse unvigintic mapping* $F : X \longrightarrow Y$ *such that*

$$\|f(x) - F(x)\| \leq \frac{L}{1 - L} \psi \left(\frac{x}{3}, \frac{x}{3} \right)$$

for all $x \in X$.

Proof The proof is obtained by similar arguments as in Theorem 6.5.1.   □

Theorem 6.5.7 *Suppose that the mapping* $f : X \longrightarrow Y$ *satisfies the inequality (6.55), for all* $x, y \in X$, *where* $\psi : X \times X \longrightarrow [0, \infty)$ *is a given function. If there exists* $L < 1$ *such that (6.48) holds for all* $x, y \in X$, *then there exists a unique multiplicative inverse unvigintic mapping* $F : X \longrightarrow Y$ *such that*

$$\|f(x) - F(x)\| \le \frac{1}{1-L} \psi\left(\frac{x}{3}, \frac{x}{3}\right)$$

for all $x \in X$.

Proof The proof is analogous to the proof of Theorem 6.5.2.   □

Corollary 6.5.8 *Let $f : X \longrightarrow Y$ be a mapping and let there exist real numbers $\alpha < -21$ and $c_1 \ge 0$ such that (6.19) holds for all $x, y \in X$. Then there exists a unique multiplicative inverse unvigintic mapping $F : X \longrightarrow Y$ satifying (6.1) and (6.20), for every $x \in X$.*

Proof The proof follows immediately by taking $\psi(x, y) = c_1 (P(x)^\alpha + P(y)^\alpha)$, for all $x, y \in X$ and $L = 3^{\alpha+21}$ in Theorem 6.5.6.   □

Corollary 6.5.9 *Let $f : X \longrightarrow Y$ be a mapping and let there exist real exist real numbers a, b such that $\rho = a + b < -21$. Let there exist $c_2 \ge 0$ such that (6.21) holds for all $x, y \in X$. Then there exists a unique multiplicative inverse unvigintic mapping $F : X \longrightarrow Y$ satisfying (6.1) and (6.22), for every $x \in X$.*

Proof The required results in Corollary 6.5.9 can be easily derived by considering $\psi(x, y) = c_2 P(x)^a P(y)^b$, for all $x, y \in X$ and $L = 3^{\rho+21}$ in Theorem 6.5.6.   □

Corollary 6.5.10 *Let $c_3 \ge 0$ and p, q be real numbers such that $\lambda = p + q < -21$, and $f : X \longrightarrow Y$ be a mapping satisfying the functional inequality (6.23) for all $x, y \in X$. Then there exists a unique multiplicative inverse unvigintic mapping $F : X \longrightarrow Y$ satisfying (6.1) and (6.24), for every $x \in X$.*

Proof By choosing $\psi(x, y) = c_3 \left(P(x)^p P(y)^q + \left(P(x)^{p+q} + P(y)^{p+q}\right)\right)$, for all $x, y \in X$ and $L = 3^{\lambda+21}$ in Theorem 6.5.6, the proof of Corollary 6.5.10 is complete.   □

6.6   Stability of (6.2): Fixed Point Method

Theorem 6.6.1 *Suppose that the mapping $f : Y \to X$ satisfies the inequality*

$$P\left(\Delta_2 f(x, y)\right) \le \psi(x, y) \tag{6.56}$$

for all $x, y \in Y$, where $\psi : Y \times Y \longrightarrow [0, \infty)$ is a given function. If there exists $L < 1$ such that

$$\psi(x, y) \le \frac{1}{3^{22}} L\psi\left(\frac{x}{3}, \frac{y}{3}\right) \tag{6.57}$$

for all $x, y \in Y$, then there exists a unique multiplicative inverse duovigintic mapping $F : Y \longrightarrow X$ such that

$$P(F(x) - f(x)) \le \frac{L}{1-L} \psi \left(\frac{x}{3}, \frac{x}{3} \right) \tag{6.58}$$

for all $x \in Y$.

Theorem 6.6.2 *Suppose that the mapping $f : Y \longrightarrow X$ satisfies the inequality (6.56), for all $x, y \in Y$, where $\psi : Y \times Y \longrightarrow [0, \infty)$ is a given function. If there exists $L < 1$ such that*

$$\psi \left(\frac{x}{3}, \frac{y}{3} \right) \le 3^{22} L \psi(x, y) \tag{6.59}$$

for all $x, y \in Y$. Then there exists a unique multiplicative inverse duovigintic mapping $F : Y \longrightarrow X$ such that

$$P(f(x) - F(x)) \le \frac{1}{1-L} \psi \left(\frac{x}{3}, \frac{x}{3} \right)$$

for all $x \in Y$.

Corollary 6.6.3 *Let $f : Y \longrightarrow X$ be a mapping and let there exist real numbers $\alpha \ne -22$ and $c_1 \ge 0$ such that (6.34) holds for all $x, y \in Y$. Then there exists a unique multiplicative inverse duovigintic mapping $F : Y \longrightarrow X$ satisfying (6.2) and (6.35), for every $x \in Y$.*

Proof The proof follows immediately by taking $\psi(x, y) = c_1 (\|x\|^\alpha + \|y\|^\alpha)$, for all $x, y \in Y$ and $L = 3^{\alpha+22}$, $L = 3^{-\alpha-22}$ in Theorems 6.6.1 and 6.6.2 respectively. □

Corollary 6.6.4 *Let $f : Y \longrightarrow X$ be a mapping and let there exist real numbers a, b such that $\rho = a + b \ne -22$. Let there exist $c_2 \ge 0$ such that (6.36) holds for all $x, y \in Y$. Then there exists a unique multiplicative inverse duovigintic mapping $F : Y \longrightarrow X$ satisfying (6.2) and (6.37), for every $x \in Y$.*

Proof The required results in Corollary 6.6.4 can be easily derived by considering $\psi(x, y) = c_2 \|x\|^a \|y\|^b$, for all $x, y \in Y$ and $L = 3^{\rho+22}$, $L = 3^{-\rho-22}$ in Theorems 6.6.1 and 6.6.2 respectively. □

Corollary 6.6.5 *Let $c_3 \ge 0$ and p, q be real numbers such that $\lambda = p + q \ne -22$, and $f : Y \longrightarrow X$ be a mapping satisfying the functional inequality (6.38), for all $x, y \in Y$. Then there exists a unique multiplicative inverse duovigintic mapping $F : Y \longrightarrow X$ satisfying (6.2) and (6.39), for every $x \in Y$.*

Proof By choosing $\psi(x, y) = c_3 \left(\|x\|^p \|y\|^q + + \left(\|x\|^{p+q} + \|y\|^{p+q} \right) \right)$, for all $x, y \in Y$ and $L = 3^{\lambda+22}$, $L = 3^{-\lambda-22}$ in Theorems 6.6.1 and 6.6.2 respectively, the proof of Corollary 6.5.5 is complete. □

Theorem 6.6.6 *Suppose that the mapping $f : X \longrightarrow Y$ satisfies the inequality*

$$\|\Delta_2 f(x, y)\| \le \psi(x, y) \tag{6.60}$$

for all $x, y \in X$, where $\psi : X \times X \longrightarrow [0, \infty)$ is a given function. If there exists $L < 1$ such that (6.57) holds for all $x, y \in X$, then there exists a unique multiplicative inverse duovigintic mapping $F : X \longrightarrow Y$ such that

$$\|f(x) - F(x)\| \leq \frac{L}{1-L}\psi\left(\frac{x}{3}, \frac{x}{3}\right)$$

for all $x \in X$.

Theorem 6.6.7 *Suppose that the mapping $f : X \longrightarrow Y$ satisfies the inequality (6.56), for all $x, y \in X$, where $\psi : X \times X \longrightarrow [0, \infty)$ is a given function. If there exists $L < 1$ such that (6.57) holds for all $x, y \in X$, then there exists a unique multiplicative inverse duovigintic mapping $F : X \longrightarrow Y$ such that*

$$\|f(x) - F(x)\| \leq \frac{1}{1-L}\psi\left(\frac{x}{3}, \frac{x}{3}\right)$$

for all $x \in X$.

Corollary 6.6.8 *Let $f : X \longrightarrow Y$ be a mapping and let there exist real numbers $\alpha < -22$ and $c_1 \geq 0$ such that (6.41) holds for all $x, y \in X$. Then there exists a unique multiplicative inverse duovigintic mapping $F : X \longrightarrow Y$ satifying (6.2) and (6.42), for every $x \in X$.*

Proof The proof follows immediately by taking $\psi(x, y) = c_1 (P(x)^\alpha + P(y)^\alpha)$, for all $x, y \in X$ and $L = 3^{\alpha+22}$ in Theorem 6.6.6. $\qquad\square$

Corollary 6.6.9 *Let $f : X \longrightarrow Y$ be a mapping and let there exist real exist real numbers a, b such that $\rho = a + b < -22$. Let there exist $c_2 \geq 0$ such that (6.43) holds for all $x, y \in X$. Then there exists a unique multiplicative inverse duovigintic mapping $F : X \longrightarrow Y$ satisfying (6.2) and (6.44), for every $x \in X$.*

Proof The required results in Corollary 6.6.9 can be easily derived by considering $\psi(x, y) = c_2 P(x)^a P(y)^b$, for all $x, y \in X$ and $L = 3^{\rho+22}$ in Theorem 6.6.6. $\qquad\square$

Corollary 6.6.10 *Let $c_3 \geq 0$ and p, q be real numbers such that $\lambda = p + q < -22$, and $f : X \longrightarrow Y$ be a mapping satisfying the functional inequality (6.45) for all $x, y \in X$. Then there exists a unique multiplicative inverse duovigintic mapping $F : X \longrightarrow Y$ satisfying (6.2) and (6.46), for every $x \in X$.*

Proof By choosing $\psi(x, y) = c_3 \left(P(x)^p P(y)^q + \left(P(x)^{p+q} + P(y)^{p+q}\right)\right)$, for all $x, y \in X$ and $L = 3^{\lambda+22}$ in Theorem 6.6.6, the proof of Corollary 6.6.10 is complete. $\qquad\square$

6.7 Counter-Examples

In this section, we show that the Eqs. (6.1) and (6.2) are not valid for $\alpha = -21$ in Corollary 6.3.3 and $\alpha = -22$ in Corollary 6.4.3, respectively, in the setting of non-zero real numbers.

Example 6.7.1 Let us define the function

$$\chi(x) = \begin{cases} \frac{c}{x^{21}}, & \text{for } x \in (1, \infty) \\ c, & \text{elsewhere} \end{cases} \tag{6.61}$$

where $\chi : \mathbb{R}^* \longrightarrow \mathbb{R}$. Let $f : \mathbb{R}^* \longrightarrow \mathbb{R}$ be a function defined as

$$f(x) = \sum_{m=0}^{\infty} 10460353203^{-m} \chi(3^{-m}x) \tag{6.62}$$

for all $x \in \mathbb{R}$. Suppose the mapping $f : \mathbb{R}^* \longrightarrow \mathbb{R}$ described in (6.62) satisfies the inequality

$$|\Delta_1 f(x, y)| \leq \frac{15690529805 \, c}{5230267601} \left(|x|^{-19} + |y|^{-19} \right) \tag{6.63}$$

for all $x, y \in \mathbb{R}^*$. We prove that there do not exist a multiplicative inverse unvigintic mapping $F : \mathbb{R}^* \longrightarrow \mathbb{R}$ and a constant $\delta > 0$ such that

$$|f(x) - F(x)| \leq \delta |x|^{-21} \tag{6.64}$$

for all $x \in \mathbb{R}^*$. Firstly, let us prove that f satisfies (6.63). Using (6.61), we have

$$|f(x)| = \left| \sum_{m=0}^{\infty} 10460353203^{-m} \chi(3^{-m}x) \right| \leq \sum_{m=0}^{\infty} \frac{c}{10460353203^m} = \frac{10460353203}{10460353202} c.$$

We find that that f is bounded by $\frac{10460353203}{10460353202} \frac{c}{}$ on $\mathbb{R}$. If $|x|^{-21} + |y|^{-21} \geq 1$, then the left hand side of (6.63) is less than $\frac{15690529805}{5230267601} \frac{c}{}$. Now, suppose that $0 < |x|^{-21} + |y|^{-21} < 1$. Hence, there exists a positive integer m such that

$$\frac{1}{10460353203^{m+1}} \leq |x|^{-21} + |y|^{-21} < \frac{1}{10460353203^m}. \tag{6.65}$$

Hence, the inequality (6.65) implies $10460353203^m \left(|x|^{-21} + |y|^{-21} \right) < 1$, or equivalently; $10460353203^m x^{-21} < 1$, $10460353203^m y^{-21} < 1$.
So, $\frac{x^{21}}{10460353203^m} > 1$, $\frac{y^{21}}{10460353203^m} > 1$. Hence, the last inequalities imply $\frac{x^{21}}{10460353203^{m-1}} > 10460353203 > 1$, $\frac{y^{21}}{10460353203^{m-1}} > 10460353203 > 1$ and as a result, we find $\frac{1}{3^{m-1}}(x) > 1$, $\frac{1}{3^{m-1}}(y) > 1$, $\frac{1}{3^{m-1}}(2x + y) > 1$, $\frac{1}{3^{m-1}}(2x - y) > 1$.

Hence, for every value of $m = 0, 1, 2, \ldots, n - 1$, we obtain

$$\frac{1}{3^n}(x) > 1, \ \frac{1}{3^n}(y) > 1, \ \frac{1}{3^n}(2x + y) > 1, \ \frac{1}{3^n}(2x - y) > 1$$

and $\Delta_1 f(3^{-n}x, 3^{-n}y) = 0$ for $m = 0, 1, 2, \ldots, n - 1$. Applying (6.61) and the definition of f, we obtain

$$|\Delta_1 f(x, y)|$$

$$\leq \sum_{m=n}^{\infty} \frac{c}{10460353203^m} + \sum_{m=n}^{\infty} \frac{c}{10460353203^m} + \frac{10460353204}{10460353203} \sum_{m=n}^{\infty} \frac{c}{10460353203^m}$$

$$\leq \frac{31381059610\,c}{10460353203} \frac{1}{10460353203^m} \left(1 - \frac{1}{10460353203}\right)^{-1}$$

$$\leq \frac{31381059610\,c}{10460353202} \frac{1}{10460353203^{m+1}}$$

$$\leq \frac{15690529805\,c}{5230267601} \left(|x|^{-21} + |y|^{-21}\right)$$

for all $x, y \in \mathbb{R}^*$. This means that the inequality (6.63) holds. We claim that the multiplicative inverse unvigintic functional equation (6.1) is unstable for $\alpha = -21$ in Corollary 6.3.3. Assume that there exists a multiplicative inverse unvigintic mapping $F : \mathbb{R}^* \longrightarrow \mathbb{R}$ satisfying (6.64). So, we have

$$|F(x)| \leq (\delta + 1)|x|^{-21}. \tag{6.66}$$

Moreover, it is possible to choose a positive integer m with the condition $mc > \delta + 1$. If $x \in (1, 3^{m-1})$, then $3^{-n}x \in (1, \infty)$ for all $m = 0, 1, 2, \ldots, n - 1$ and thus

$$|F(x)| = \sum_{m=0}^{\infty} \frac{\chi(3^{-m}x)}{10460353203^m} \geq \sum_{m=0}^{n-1} \frac{\frac{10460353203^m c}{x^{21}}}{10460353203^m} = \frac{mc}{x^{21}} > (\delta + 1)x^{-21}$$

which contradicts (6.66). Therefore, the multiplicative inverse unvigintic functional equation (6.1) is unstable for $\alpha = -21$ in Corollary 6.3.3.

Similar to Example 6.7.1, the following example acts as a counter-example that the Eq. (6.2) is not stable for $\alpha = -22$ in Corollary 6.4.3.

Example 6.7.2 Define the function $\xi : \mathbb{R}^* \longrightarrow \mathbb{R}$ via

$$\xi(x) = \begin{cases} \frac{\lambda}{x^{22}} & \text{for } u \in (1, \infty) \\ c, & \text{otherwise} \end{cases}.$$

Let $f : \mathbb{R}^* \longrightarrow \mathbb{R}$ be defined by

$$f(x) = \sum_{m=0}^{\infty} 31381059609^{-m} \xi(3^{-m}x)$$

for all $x \in \mathbb{R}$. Suppose the function f satisfies the inequality

$$|\Delta_2 f(x, y)| \leq \frac{47071589414 \, \lambda}{15690529804} \left(|x|^{-22} + |y|^{-22}\right)$$

for all $x, y \in \mathbb{R}^*$. Then, there do not exist a multiplicative inverse duovigintic mapping $F : \mathbb{R}^* \longrightarrow \mathbb{R}$ and a constant $\eta > 0$ such that

$$|f(x) - F(x)| \leq \eta \, |x|^{-22}$$

for all $x \in \mathbb{R}^*$.

Chapter 7
Inexact Solution of Multiplicative Inverse Type Trevigintic and Quottuorvigintic Functional Equations in Matrix Normed Spaces

Abstract In this chapter, an inexact solution near to the exact solution of a multiplicative inverse trevigintic and quottuorvigintic functional equations are achieved in the sense of Ulam stability hypothesis in matrix normed spaces. Proper examples are also illustrated to prove the instabilities for control cases.

7.1 Introduction

The quotient spaces, mapping spaces and other tensor products of operator spaces may be considered as operator spaces due to the abstract characterization provided for linear spaces of bounded Hilbert spaces operators in terms of matrix normed spaces [125]. In lieu of this result, there is a lot of significant scope for the theory of operator spaces in operator algebra theory (see [29]). The proof provided in [125] claimed to the hypothesis of ordered operator spaces [20]. By employing the technique established in [107], one can provide a purely metric proof of this result [30] (corrected version of in [27]). The Hyers-Ulam stabilities of Cauchy and quadratic functional equations were dealt in [74].

Here, we present the concepts of matrix normed spaces, then we discuss the Ulam-Hyers stability of multiplicative inverse trevigintic and quottuorvigintic functional equations in matrix normed spaces through various theorems. We will use the following notations:

$M_n(X)$ is the set of all $n \times n$-matrices in X;

$e_j \in M_{1,n}(\mathbb{C})$ is that jth component is 1, and the other components are zero;

$E_{ij} \in M_n(\mathbb{C})$ is that (i, j)-component is 1, and the other components are zero;

$E_{ij} \otimes x \in M_n(X)$ is that (i, j)-component is x, and the other components are zero.

For $x \in M_n(X), y \in M_k(X)$,

$$x \oplus y = \begin{pmatrix} x & 0 \\ 0 & y \end{pmatrix}.$$

© The Editor(s) (if applicable) and The Author(s), under exclusive license
to Springer Nature Switzerland AG 2020
B. V. Senthil Kumar and H. Dutta, *Multiplicative Inverse Functional Equations*,
Studies in Systems, Decision and Control 289, https://doi.org/10.1007/978-3-030-45355-8_7

Definition 7.1.1 Let $(X, || \cdot ||)$ be a normed space. Note that $(X, || \cdot ||_n)$ is a matrix normed space if and only if $(M_n(X), || \cdot ||_n)$ is a normed space for each positive integer n and $||AxB||_k \leq ||A|| \, ||B|| \, ||x||_n$ holds for $A \in M_{k,n}(\mathbb{C})$, $x = (x_{ij}) \in M_n(X)$ and $B \in M_{n,k}(\mathbb{C})$.

Definition 7.1.2 $(X, || \cdot ||_n)$ is a matrix Banach space if and only if x is a Banach space and $(X, || \cdot ||_n)$ is matrix normed space.

Definition 7.1.3 A matrix normed space $(X, || \cdot ||_n)$ is called and L^∞-*matrix normed space* if $||x \oplus y||_{n+k} = \max\{||x||_n, ||y||_k\}$ holds for all $x \in M_n(X)$ and all $y \in M_k(X)$.

Let E, F be vector spaces. For a given mapping $h : E \to F$ and a given positive integer n, define $h_n : M_n(E) \to M_n(F)$ by

$$h_n \left([x_{ij}] \right) = \left[h \left(x_{ij} \right) \right]$$

for all $\left[x_{ij} \right] \in M_n(E)$.

In this chapter, we deal with the multiplicative inverse trevigintic functional equation

$$f(2a + b) + f(2a - b)$$
$$= \frac{4f(a)f(b)}{\left(4f(b)^{2/23} - f(a)^{2/23}\right)^{23}} \left[\frac{1}{2} \sum_{k=0}^{11} \binom{23}{2k} [f(a)]^{2k/23}[f(b)]^{(23-2k)/23} \right] \quad (7.1)$$

and a multiplicative inverse quottuorvigintic functional equation

$$f(2a + b) + f(2a - b)$$
$$= \frac{4f(a)f(b)}{\left(4f(b)^{1/12} - f(a)^{1/12}\right)^{24}} \left[\sum_{k=0}^{12} \binom{24}{2k} [f(a)]^{k/12}[f(b)]^{(24-2k)/24} \right]. \quad (7.2)$$

We investigate the generalized Hyers-Ulam stability of the functional equations (7.1) and (7.2) in matrix normed spaces. We also prove that the stability results are false through proper counter-examples for critical cases.

7.2 Stability of Eq. (7.1) in matrix normed spaces

In the following theorems, let $(X, || \cdot ||_n)$ be a matrix normed space and $(Y, || \cdot ||_n)$ be a matrix Banach space. The generalized-Ulam-Hyers stability of the multiplicative inverse trevigintic functional equation (7.1) in matrix normed spaces is presented in the following theorem. For a mapping $f : X \longrightarrow Y$, define $\Delta_1 f : X \times X \longrightarrow Y$ and $\Delta_1 f_n : M_n(X \times X) \longrightarrow M_n(Y)$ by

$\Delta_1 f(a, b) = f(2a + b) + f(2a - b)$

$$- \frac{4f(a)f(b)}{\left(4f(b)^{2/23} - f(a)^{2/23}\right)^{23}} \left[\frac{1}{2} \sum_{k=0}^{11} \binom{23}{2k} [f(a)]^{2k/23} [f(b)]^{(23-2k)/23} \right],$$

$\Delta_1 f_n \left([x_{ij}], [y_{ij}]\right)$
$$= f_n(2[x_{ij}] + [y_{ij}]) + f_n(2[x_{ij}] - [y_{ij}])$$

$$- \frac{4f_n([x_{ij}])f([y_{ij}])}{\left(4f([y_{ij}])^{2/23} - f([x_{ij}])^{2/23}\right)^{23}} \left[\frac{1}{2} \sum_{k=0}^{11} \binom{23}{2k} [f([x_{ij}])]^{2k/23} [f([y_{ij}])]^{(23-2k)/23} \right]$$

for all $a, b \in X$, and all $x = [x_{ij}], y = [y_{ij}] \in M_n(X)$.

Lemma 7.2.1 ([28]) *Let $(X, \|\cdot\|_n)$ be a matrix normed space. Then*

(i) $\|E_{kl} \otimes x\|_n = \|x\|$ *for $x \in X$.*
(ii) $\|x_{kl}\| \le \|[x_{ij}]\|_n \le \sum_{i,j=1}^n \|x_{ij}\|$ *for $[x_{ij}] \in M_n(X)$.*
(iii) $\lim_{n\to\infty} x_n = x$ *if and only if* $\lim_{n\to\infty} x_{nij} = x_{ij}$ *for $x_n = [x_{nij}], x = [x_{ij}] \in M_k(X)$.*

Proof (i) Since $E_{kl} \otimes x = \dot{e}_k x \dot{e}_l$ and $\|\dot{e}_k\| = \|\dot{e}_l\| = 1$, $\|E_{kl} \otimes x\|_n \le \|x\|$. Since $e_k (E_{kl} \otimes x) \dot{e}_l = x$, $\|x\| \le \|E_{kl} \otimes x\|_n$. So, $\|E_{kl} \otimes x\|_n = \|x\|$.
(ii) Since $e_k x \dot{e}_l = x_{kl}$ and $\|e_k\| = \|\dot{e}_l\| = 1$, we have

$$\|x_{kl}\| \le \|[x_{ij}]\|_n.$$

Since $[x_{ij}] = \sum_{i,j=1}^n E_{ij} \otimes x_{ij}$, we obtain

$$\|[x_{ij}]\|_n = \left\| \sum_{i,j=1}^n E_{ij} \otimes x_{ij} \right\|_n \le \sum_{i,j=1}^n \|E_{ij} \otimes x_{ij}\|_n = \sum_{i,j=1}^n \|x_{ij}\|.$$

(iii) By (ii), we have

$$\|x_{nkl} - x_{kl}\| \le \|[x_{nij} - x_{ij}]\|_n = \|[x_{nij}] - [x_{ij}]\|_n \le \sum_{i,j=1}^n \|x_{nij} - x_{ij}\|.$$

Hence, we get the desired result. This completes the proof. $\square$

Theorem 7.2.2 *Let $f : X \longrightarrow Y$ be a mapping and let $\phi : X \times X \longrightarrow [0, \infty)$ be a function such that*

$$\Phi(a, b) = \frac{1}{3} \sum_{l=0}^{\infty} \frac{1}{3^{23l}} \phi \left(\frac{a}{3^l}, \frac{b}{3^l} \right) < +\infty, \tag{7.3}$$

$$\left\| \Delta_1 f_n \left([x_{ij}], [y_{ij}]\right) \right\|_n \leq \sum_{i,j=1}^{n} \phi(x_{ij}, y_{ij}) \tag{7.4}$$

for all $a, b \in X$, and all $x = [x_{ij}]$, $y = [y_{ij}] \in M_n(X)$. Then there exists a unique multiplicative inverse trevigintic mapping $F : X \longrightarrow Y$ such that

$$\left\| f_n([x_{ij}]) - F_n([x_{ij}]) \right\|_n \leq \sum_{i,j=1}^{n} \Phi(x_{ij}, x_{ij}) \tag{7.5}$$

for all $x = [x_{ij}] \in M_n(X)$.

Proof Assume $n = 1$ in (7.4). Then we have

$$\|\Delta_1 f(a, b)\| \leq \phi(a, b)$$

for all $a, b \in X$. Then there exists a unique multiplicative inverse trevigintic mapping $F : X \longrightarrow Y$ such that

$$\|f(a) - F(a)\| \leq \Phi(a, a)$$

for all $a \in X$. Define the mapping $F : X \longrightarrow Y$ by

$$F(a) = \lim_{l \to \infty} \frac{1}{3^{23l}} f\left(\frac{a}{3^l}\right)$$

for all $a \in X$. In view of Lemma 7.2.1, we have

$$\left\| f_n([x_{ij}]) - F_n([x_{ij}]) \right\|_n \leq \sum_{i,j=1}^{n} \left\| f(x_{ij}) - F(x_{ij}) \right\| \leq \sum_{i,j=1}^{n} \Phi(x_{ij}, x_{ij})$$

for all $x = [x_{ij}] \in M_n(X)$. Thus, $F : X \longrightarrow Y$ is a unique multiplicative inverse trevigintic mapping satisfying (7.5), as desired. The proof of the theorem is completed. $\qquad\square$

Corollary 7.2.3 *Let r, θ be positive real numbers with $r < -23$. Let $f : X \longrightarrow Y$ be a mapping such that*

$$\left\| Df_n([x_{ij}], [y_{ij}]) \right\|_n \leq \sum_{i,j=1}^{n} \theta \left(\|x_{ij}\|^r + \|y_{ij}\|^r \right) \tag{7.6}$$

for all $x = [x_{ij}]$, $y = [y_{ij}] \in M_n(X)$. Then there exists a unique multiplicative inverse trevigintic mapping $F : X \longrightarrow Y$ such that

$$\left\| f_n([x_{ij}]) - F_n([x_{ij}]) \right\|_n \leq \sum_{i,j=1}^{n} \frac{2\theta}{3^{-23} - 3^r} \|x_{ij}\|^r$$

for all $x = [x_{ij}] \in M_n(X)$.

Proof Letting $\phi(a,b) = \theta\left(\|a\|^r + \|b\|^r\right)$ in Theorem 7.2.2, we obtain the result, and the proof is completed.   □

Theorem 7.2.4 *Let $f : X \longrightarrow Y$ be a mapping and let $\phi : X \times X \longrightarrow [0, \infty)$ be a function satisfying (7.4), and*

$$\Phi(a,b) = \frac{1}{3} \sum_{l=1}^{\infty} 3^{23l} \phi\left(3^l a, 3^l b\right) < +\infty \tag{7.7}$$

for all $a, b \in X$. Then there exists a unique multiplicative inverse treviginic mapping $F : X \longrightarrow Y$ such that

$$\left\| f_n([x_{ij}]) - F_n([x_{ij}]) \right\|_n \leq \sum_{i,j=1}^{n} \Phi(x_{ij}, x_{ij})$$

for all $x = [x_{ij}] \in M_n(X)$.

Proof The proof is similar to that of Theorem 7.2.2, and hence it is omitted.   □

Corollary 7.2.5 *Let r, θ be positive real numbers with $r > -23$. Let $f : X \longrightarrow Y$ be a mapping satisfying (7.6). Then there exists a unique multiplicative inverse treviginic mapping $F : X \longrightarrow Y$ such that*

$$\left\| f_n([x_{ij}]) - F_n([x_{ij}]) \right\|_n \leq \sum_{i,j=1}^{n} \frac{2\theta}{3^r - 3^{-23}} \|x_{ij}\|^r$$

for all $x = [x_{ij}] \in M_n(X)$.

Proof Letting $\phi(a,b) = \theta\left(\|a\|^r + \|b\|^r\right)$ in Theorem 7.2.4, we obtain the result. This completes the proof.   □

We need the following lemma to prove our main results.

Lemma 7.2.6 *[136]. If E is a L^∞-matrix normed space, then $\left\|[x_{ij}]\right\|_n \leq \left\|[\|x_{ij}\|]\right\|_n$ for all $[x_{ij}] \in M_n(E)$.*

Theorem 7.2.7 *Let Y be a L^∞-normed Banach space. Let $f : X \longrightarrow Y$ be a mapping and let $\phi : X \times X \longrightarrow [0, \infty)$ be a function satisfying (7.3), and*

$$\left\| \Delta_1 f_n([x_{ij}], [y_{ij}]) \right\|_n \leq \left\|[\phi(x_{ij}, y_{ij})]\right\|_n \tag{7.8}$$

for all $x = [x_{ij}]$, $y = [y_{ij}] \in M_n(X)$. *Then there exists a unique multiplicative inverse trevigintic mapping* $F : X \longrightarrow Y$ *such that*

$$\left\| [f(x_{ij}) - F(x_{ij})] \right\|_n \leq \left\| [\Phi(x_{ij}, x_{ij})] \right\|_n \tag{7.9}$$

for all $x = [x_{ij}] \in M_n(X)$. *Here* Φ *is given in Theorem 7.2.2.*

Proof By the same reasoning as in the proof of Theorem 7.2.2, there exists a unique multiplicative inverse trevigintic mapping $F : X \longrightarrow Y$ such that

$$\| f(a) - F(a) \| \leq \Phi(a, a)$$

for all $a \in X$. The mapping $F : X \longrightarrow Y$ is given by

$$F(a) = \lim_{l \to \infty} \frac{1}{3^{23l}} f\left(\frac{a}{3^l}\right)$$

for all $a \in X$.

It is easy to show that if $0 \leq a_{ij} \leq b_{ij}$ for all i, j, then

$$\left\| [a_{ij}] \right\|_n \leq \left\| [b_{ij}] \right\|_n . \tag{7.10}$$

By Lemma 7.2.6 and (7.10), we have

$$\left\| [f(x_{ij}) - F(x_{ij})] \right\|_n \leq \left\| [\|f(x_{ij}) - F(x_{ij})\|] \right\|_n \leq \left\| \Phi(x_{ij}, x_{ij}) \right\|_n$$

for all $x = [x_{ij}] \in M_n(X)$. So, we obtain the inequality (7.9). This completes the proof. $\qquad \square$

Corollary 7.2.8 *Let* Y *be a* L^∞-*normed Banach space. Let* r, θ *be positive real numbers with* $r < -23$. *Let* $f : X \longrightarrow Y$ *be a mapping such that*

$$\left\| \Delta_1 f_n([x_{ij}], [y_{ij}]) \right\|_n \leq \left\| [\theta \left(\|x_{ij}\|^r + \|y_{ij}\|^r \right)] \right\|_n \tag{7.11}$$

for all $x = [x_{ij}]$, $y = [y_{ij}] \in M_n(X)$. *Then there exists a unique multiplicative inverse trevigintic mapping* $F : X \longrightarrow Y$ *such that*

$$\left\| f_n([x_{ij}]) - F_n([x_{ij}]) \right\|_n \leq \left\| \left[\frac{2\theta}{3^{-23} - 3^r} \|x_{ij}\|^r \right] \right\|_n$$

for all $x = [x_{ij}] \in M_n(X)$.

Proof Letting $\phi(a, b) = \theta \left(\|a\|^r + \|b\|^r \right)$ in Theorem 7.2.7, we obtain the desired result. This completes the proof. $\qquad \square$

Theorem 7.2.9 *Let Y be a L^∞-normed Banach space. Let $f : X \longrightarrow Y$ be a mapping and let $\phi : X \times X \longrightarrow [0, \infty)$ be a function satisfying (7.7), and (7.8). Then there exists a unique multiplicative inverse trevigintic mapping $F : X \longrightarrow Y$ such that*

$$\left\| [f(x_{ij}) - F(x_{ij})] \right\|_n \leq \left\| [\Phi(x_{ij}, x_{ij})] \right\|_n$$

for all $x = [x_{ij}] \in M_n(X)$. Here Φ is given in Theorem 7.2.4.

Proof The proof is similar to Theorem 7.2.7, and hence the details are omitted. □

Corollary 7.2.10 *Let Y be a L^∞-normed Banach space. Let r, θ be positive real numbers with $r > -23$. Let $f : X \longrightarrow Y$ be a mapping satisfying (7.11). Then there exists a unique multiplicative inverse trevigintic mapping $F : X \longrightarrow Y$ such that*

$$\left\| f_n([x_{ij}] - F_n([x_{ij}])) \right\|_n \leq \left\| \left[\frac{2\theta}{3^r - 3^{-23}} \|x_{ij}\|^r \right] \right\|_n$$

for all $x = [x_{ij}] \in M_n(X)$.

Proof Letting $\phi(a, b) = \theta \left(\|a\|^r + \|b\|^r \right)$ in Theorem 7.2.9, we obtain the desired result. This completes the proof.   □

7.3   Stability of Eq. (7.2) in matrix normed spaces

In the following theorems, we present the generalized Ulam-Hyers stability of the multiplicative inverse quottuorvigintic functional equation (7.2) in matrix normed spaces. For a mapping $f : X \longrightarrow Y$, define $\Delta_2 f : X \times X \longrightarrow Y$ and $\Delta_2 f_n : M_n(X \times X) \to M_n(Y)$ by

$$\Delta_2 f(a, b) = f(2a + b) + f(2a - b)$$
$$- \frac{4f(a)f(b)}{\left(4f(b)^{1/12} - f(a)^{1/12}\right)^{24}} \left[\sum_{k=0}^{12} \binom{24}{2k} [f(a)]^{k/12} [f(b)]^{(24-2k)/24} \right],$$

and

$$\Delta_2 f_n \left([x_{ij}], [y_{ij}] \right)$$
$$= f \left(2[x_{ij}] + [y_{ij}] \right) + f \left(2[x_{ij}] - [y_{ij}] \right)$$
$$- \frac{4f \left([x_{ij}] \right) f \left([y_{ij}] \right)}{\left(4f \left([y_{ij}] \right)^{1/12} - f \left([x_{ij}] \right)^{1/12} \right)^{24}} \left[\sum_{k=0}^{12} \binom{24}{2k} [f \left([x_{ij}] \right)]^{k/12} [f \left([y_{ij}] \right)]^{(24-2k)/24} \right],$$

for all $a, b \in X$, and all $x = [x_{ij}], y = [y_{ij}] \in M_n(X)$.

Theorem 7.3.1 *Let $f : X \longrightarrow Y$ be a mapping, and let $\phi : X \times X \longrightarrow [0, \infty)$ be a function such that*

$$\Phi(a, b) = \frac{1}{3} \sum_{l=0}^{\infty} \frac{1}{3^{24l}} \phi \left(\frac{a}{3^l}, \frac{b}{3^l} \right) < +\infty, \tag{7.12}$$

and

$$\left\| Df_n([x_{ij}], [y_{ij}]) \right\|_n \leq \sum_{i,j=1}^{n} \phi(x_{ij}, y_{ij}) \tag{7.13}$$

for all $a, b \in X$, and all $x = [x_{ij}]$, $y = [y_{ij}] \in M_n(X)$. Then there exists a unique multiplicative inverse quottuorvigintic mapping $F : X \longrightarrow Y$ such that

$$\left\| f_n([x_{ij}]) - F_n([x_{ij}]) \right\|_n \leq \sum_{i,j=1}^{n} \Phi(x_{ij}, x_{ij}) \tag{7.14}$$

for all $x = [x_{ij}] \in M_n(X)$.

Proof Let $n = 1$ in (7.13). Then, we have $\|\Delta_2 f(a, b)\| \leq \phi(a, b)$, for all $a, b \in X$. Then there exists a unique multiplicative inverse quottuovigintic mapping $F : X \longrightarrow Y$ such that

$$\|f(a) - F(a)\| \leq \Phi(a, a)$$

for all $a \in X$. Define $F : X \longrightarrow Y$ by

$$F(a) = \lim_{l \to \infty} \frac{1}{3^{24l}} f \left(\frac{a}{3^l} \right)$$

for all $a \in X$. Then by Lemma 7.2.1, we have

$$\left\| f_n([x_{ij}]) - F_n([x_{ij}]) \right\|_n \leq \sum_{i,j=1}^{n} \left\| f(x_{ij}) - F(x_{ij}) \right\|$$

$$\leq \sum_{i,j=1}^{n} \Phi(x_{ij}, x_{ij})$$

for all $x = [x_{ij}] \in M_n(X)$. Thus, $F : X \longrightarrow Y$ is a unique multiplicative inverse quottuorvigintic mapping satisfying (7.14), as desired. This completes the proof. $\Box$

Corollary 7.3.2 *Let r, θ be positive real numbers with $r < -24$. Let $f : X \longrightarrow Y$ be a mapping such that*

$$\left\| \Delta_2 f_n([x_{ij}], [y_{ij}]) \right\|_n \leq \sum_{i,j=1}^{n} \theta \left(\|x_{ij}\|^r + \|y_{ij}\|^r \right) \tag{7.15}$$

for all $x = [x_{ij}]$, $y = [y_{ij}] \in M_n(X)$. *Then there exists a unique multiplicative inverse quottuorvigintic mapping* $F : X \longrightarrow Y$ *such that*

$$\left\| f_n([x_{ij}]) - F_n([x_{ij}]) \right\|_n \leq \sum_{i,j=1}^{n} \frac{2\theta}{3^{-24} - 3^r} \|x_{ij}\|^r$$

for all $x = [x_{ij}] \in M_n(X)$.

Proof Letting $\phi(a, b) = \theta \left(\|a\|^r + \|b\|^r \right)$ in Theorem 7.3.1, we obtain the result. The proof is now completed. □

Theorem 7.3.3 *Let* $f : X \longrightarrow Y$ *be a mapping and let* $\phi : X \times X \longrightarrow [0, \infty)$ *be a function satisfying (7.13), and*

$$\Phi(a, b) = \frac{1}{3} \sum_{l=1}^{\infty} 3^{24l} \phi \left(3^l a, 3^l b \right) < +\infty \qquad (7.16)$$

for all $a, b \in X$. *Then there exists a unique multiplicative inverse quottuorvigintic mapping* $F : X \longrightarrow Y$ *such that*

$$\left\| f_n([x_{ij}]) - F_n[(x_{ij})] \right\|_n \leq \sum_{i,j=1}^{n} \Phi(x_{ij}, x_{ij})$$

for all $x = [x_{ij}] \in M_n(X)$.

Proof The proof is similar to Theorem 7.3.1, and the details are left to the reader. □

Corollary 7.3.4 *Let* r, θ *be positive real numbers with* $r > -24$. *Let* $f : X \longrightarrow Y$ *be a mapping satisfying (7.15). Then there exists a unique multiplicative inverse quottuorvigintic mapping* $F : X \longrightarrow Y$ *such that*

$$\left\| f_n([x_{ij}]) - F_n([x_{ij}]) \right\|_n \leq \sum_{i,j=1}^{n} \frac{2\theta}{3^r - 3^{-24}} \|x_{ij}\|^r$$

for all $x = [x_{ij}] \in M_n(X)$.

Proof Letting $\phi(a, b) = \theta \left(\|a\|^r + \|b\|^r \right)$ in Theorem 7.3.3, then we obtain the desired result. This completes the proof. □

Theorem 7.3.5 *Let* $f : X \longrightarrow Y$ *be a mapping and let* $\phi : X \times X \longrightarrow [0, \infty)$ *be a function satisfying (7.12), and*

$$\left\| \Delta_2 f_n([x_{ij}], [y_{ij}]) \right\|_n \leq \left\| [\phi(x_{ij}, y_{ij})] \right\|_n \qquad (7.17)$$

for all $x = [x_{ij}]$, $y = [y_{ij}] \in M_n(X)$. *Then there exists a unique multiplicative inverse quottuorvigintic mapping* $F : X \longrightarrow Y$ *such that*

$$\left\| [f(x_{ij}) - F(x_{ij})] \right\|_n \leq \left\| \Phi(x_{ij}, x_{ij}) \right\|_n \qquad (7.18)$$

for all $x = [x_{ij}] \in M_n(X)$. Here Φ is as given in Theorem 7.3.1.

Proof By the same reasoning as in the proof of Theorem 7.3.1, there exists a unique multiplicative inverse quottuorvigintic $F : X \longrightarrow Y$ such that

$$\| f(a) - F(a) \| \leq \Phi(a, a)$$

for all $a \in X$. Define $F : X \longrightarrow Y$ by

$$F(a) = \lim_{l \to \infty} \frac{1}{3^{24l}} f\left(\frac{a}{3^l}\right)$$

for all $a \in X$, then by Lemma 7.2.1, and (7.10), we obtain

$$\| [f(x_{ij}) - F(x_{ij})] \|_n \leq \left\| [\| f(x_{ij}) - F(x_{ij}) \|] \right\|_n \leq \left\| [\Phi(x_{ij}, x_{ij})] \right\|_n$$

for all $x = [x_{ij}] \in M_n(X)$. So, we obtain the inequality (7.18). The proof is now completed. $\qquad \square$

Corollary 7.3.6 *Let r, θ be positive real numbers with $r < -24$. Let $f : X \longrightarrow Y$ be a mapping such that*

$$\left\| \Delta_2 f_n([x_{ij}], [y_{ij}]) \right\|_n \leq \left\| [\theta (\|x_{ij}\|^r + \|y_{ij}\|^r)] \right\|_n \qquad (7.19)$$

for all $x = [x_{ij}]$, $y = [y_{ij}] \in M_n(X)$. Then there exists a unique multiplicative inverse quottuorvigintic mapping $F : X \longrightarrow Y$ such that

$$\left\| f_n([x_{ij}]) - F_n([x_{ij}]) \right\|_n \leq \left\| \left[\frac{2\theta}{3^{-24} - 3^r} \|x_{ij}\|^r \right] \right\|_n$$

for all $x = [x_{ij}] \in M_n(X)$.

Proof Letting $\phi(a, b) = \theta (\|a\|^r + \|b\|^r)$ in Theorem 7.3.5, we obtain the desired result. The proof is now completed. $\qquad \square$

Theorem 7.3.7 *Let $f : X \longrightarrow Y$ be a mapping and let $\phi : X \times X \longrightarrow [0, \infty)$ be a function satisfying (7.16), and (7.17). Then there exists a unique multiplicative inverse quottuorvigintic mapping $F : X \longrightarrow Y$ such that*

$$\| [f(x_{ij}) - F(x_{ij})] \|_n \leq \left\| [\Phi(x_{ij}, x_{ij})] \right\|_n$$

for all $x = [x_{ij}] \in M_n(X)$. Here Φ is as given in Theorem 7.3.1.

Proof The proof is similar to Theorem 7.3.5, and the details are left to the reader. $\square$

Corollary 7.3.8 *Let r, θ be positive real numbers with $r > -24$. Let $f : X \longrightarrow Y$ be a mapping satisfying (7.19). Then there exists a unique multiplicative inverse quottuorvigintic mapping $F : X \longrightarrow Y$ such that*

$$\left\| f_n([x_{ij}]) - F_n([x_{ij}]) \right\|_n \leq \left\| \left[\frac{2\theta}{3^r - 2^{-24}} ||x_{ij}||^r \right] \right\|_n$$

for all $x = [x_{ij}] \in M_n(X)$.

Proof Letting $\phi(a, b) = \theta \left(||a||^r + ||b||^r \right)$ in Theorem 7.3.7, we obtain the desired result. This completes the proof.   □

7.4 Counter-Examples

In this section, we show that the Eqs. (7.1) and (7.2) are not valid for $r = -23$ in Corollary 7.2.3 and $r = -24$ in Corollary 7.3.2, respectively, in the setting of non-zero real numbers.

Example 7.4.1 Let us define the function

$$\chi(x) = \begin{cases} \frac{c}{x^{23}}, & \text{for } x \in (1, \infty) \\ c, & \text{elsewhere} \end{cases} \tag{7.20}$$

where $\chi : \mathbb{R}^* \longrightarrow \mathbb{R}$. Let $f : \mathbb{R}^* \longrightarrow \mathbb{R}$ be a function defined as

$$f(x) = \sum_{m=0}^{\infty} 94143178827^{-m} \chi(3^{-m} x) \tag{7.21}$$

for all $x \in \mathbb{R}$. Suppose the mapping $f : \mathbb{R}^* \longrightarrow \mathbb{R}$ described in (7.21) satisfies the inequality

$$|\Delta_1 f(x, y)| \leq \frac{141214768241\, c}{47071589413} \left(|x|^{-23} + |y|^{-23} \right) \tag{7.22}$$

for all $x, y \in \mathbb{R}^*$. We prove that there do not exist a multiplicative inverse trevigintic mapping $F : \mathbb{R}^* \longrightarrow \mathbb{R}$ and a constant $\delta > 0$ such that

$$|f(x) - F(x)| \leq \delta |x|^{-23} \tag{7.23}$$

for all $x \in \mathbb{R}^*$. Firstly, let us prove that f satisfies (7.22). Using (7.20), we have

$$|f(x)| = \left| \sum_{m=0}^{\infty} 94143178827^{-m} \chi(3^{-m} x) \right| \leq \sum_{m=0}^{\infty} \frac{c}{94143178827^m} = \frac{94143178827}{94143178826} c.$$

We find that that f is bounded by $\frac{94143178827}{94143178826} c$ on $\mathbb{R}$. If $|x|^{-23} + |y|^{-23} \geq 1$, then the left hand side of (7.22) is less than $\frac{141214768241}{47071589413} c$. Now, suppose that $0 < |x|^{-23} + |y|^{-23} < 1$. Hence, there exists a positive integer m such that

$$\frac{1}{94143178827^{m+1}} \leq |x|^{-23} + |y|^{-23} < \frac{1}{94143178827^{m}}. \tag{7.24}$$

Hence, the inequality (7.24) implies $94143178827^{m} \left(|x|^{-23} + |y|^{-23} \right) < 1$, or equivalently; $94143178827^{m} x^{-23} < 1, 94143178827^{m} y^{-23} < 1$.
So, $\frac{x^{23}}{94143178827^{m}} > 1, \frac{y^{23}}{94143178827^{m}} > 1$. Hence, the last inequalities imply $\frac{x^{23}}{94143178827^{m-1}} > 94143178827 > 1, \frac{y^{23}}{94143178827^{m-1}} > 94143178827 > 1$ and as a result, we find $\frac{1}{3^{m-1}}(x) > 1, \frac{1}{3^{m-1}}(y) > 1, \frac{1}{3^{m-1}}(2x + y) > 1, \frac{1}{3^{m-1}}(2x - y) > 1$. Hence, for every value of $m = 0, 1, 2, \ldots, n - 1$, we obtain

$$\frac{1}{3^{n}}(x) > 1, \frac{1}{3^{n}}(y) > 1, \frac{1}{3^{n}}(2x + y) > 1, \frac{1}{3^{n}}(2x - y) > 1$$

and $\Delta_1 f(3^{-n} x, 3^{-n} y) = 0$ for $m = 0, 1, 2, \ldots, n - 1$. Applying (7.20) and the definition of f, we obtain

$|\Delta_1 f(x, y)|$

$$\leq \sum_{m=n}^{\infty} \frac{c}{94143178827^{m}} + \sum_{m=n}^{\infty} \frac{c}{94143178827^{m}} + \frac{94143178828}{94143178827} \sum_{m=n}^{\infty} \frac{c}{94143178827^{m}}$$

$$\leq \frac{282429536482 \, c}{94143178827} \frac{1}{94143178827^{m}} \left(1 - \frac{1}{94143178827} \right)^{-1}$$

$$\leq \frac{282428536482 \, c}{94143178826} \frac{1}{94143178827^{m+1}}$$

$$\leq \frac{141214768241 \, c}{47071589413} \left(|x|^{-23} + |y|^{-23} \right)$$

for all $x, y \in \mathbb{R}^*$. This means that the inequality (7.22) holds. We claim that the multiplicative inverse trevigintic functional equation (7.1) is unstable for $r = -23$ in Corollary 7.2.3. Assume that there exists a multiplicative inverse trevigintic mapping $F : \mathbb{R}^* \longrightarrow \mathbb{R}$ satisfying (7.22). So, we have

$$|F(x)| \leq (\delta + 1)|x|^{-23}. \tag{7.25}$$

Moreover, it is possible to choose a positive integer m with the condition $mc > \delta + 1$. If $x \in \left(1, 3^{m-1} \right)$, then $3^{-n} x \in (1, \infty)$ for all $m = 0, 1, 2, \ldots, n - 1$ and thus

$$|F(x)| = \sum_{m=0}^{\infty} \frac{\chi(3^{-m} x)}{94143178827^{m}} \geq \sum_{m=0}^{n-1} \frac{\frac{94143178827^{m} c}{x^{23}}}{94143178827^{m}} = \frac{mc}{x^{23}} > (\delta + 1)x^{-23}$$

which contradicts (7.25). Therefore, the multiplicative inverse trevigintic functional equation (7.1) is unstable for $r = -23$ in Corollary 7.2.3.

Similar to Example 7.4.1, the following example acts as a counter-example that the equation (7.2) is not stable for $r = -24$ in Corollary 7.3.2.

Example 7.4.2 Define the function $\xi : \mathbb{R}^* \longrightarrow \mathbb{R}$ via

$$\xi(x) = \begin{cases} \frac{\lambda}{x^{24}} & \text{for } u \in (1, \infty) \\ c, & \text{otherwise} \end{cases} .$$

Let $f : \mathbb{R}^* \longrightarrow \mathbb{R}$ be defined by

$$f(x) = \sum_{m=0}^{\infty} 282429536481^{-m} \xi(3^{-m} x)$$

for all $x \in \mathbb{R}$. Suppose the function f satisfies the inequality

$$|\Delta_2 f(x, y)| \le \frac{423644304772 \, \lambda}{141214768240} \left(|x|^{-24} + |y|^{-24} \right)$$

for all $x, y \in \mathbb{R}^*$. Then, there do not exist a multiplicative inverse quottuorvigintic mapping $F : \mathbb{R}^* \longrightarrow \mathbb{R}$ and a constant $\eta > 0$ such that

$$|f(x) - F(x)| \le \eta \, |x|^{-24}$$

for all $x \in \mathbb{R}^*$.

References

1. Adam, M.: On the stability of some quadratic functional equation. J. Nonlinear Sci. Appl. **4**(1), 50–59 (2011)
2. Almahalebi, M., Chahbi, A., Kabbaj, S.: A fixed point approach to the stability of a bi-cubic functional equation in 2-Banach spaces. Palest. J. Math. **5**(2), 220–227 (2016)
3. Alshybani, S., Vaezpour, S.M., Saadati, R.: Stability of the sextic functional equation in various spaces. J. Inequal. Spec. Funct. **9**(4), 8–27 (2018)
4. Aoki, T.: On the stability of the linear transformation in Banach spaces. J. Math. Soc. Jpn. **2**, 64–66 (1950)
5. Arunkumar, M.: Generalized Ulam-Hyers stability of derivations of an AQ-functional equation. CUBO Math. J. **15**(1), 159–169 (2013)
6. Arunkumar, M., Karthikeyan, S.: Solution and stability of a reciprocal functional equation originating from n-consecutive terms of a harmonic progression: direct and fixed point methods. Int. J. Inf. Sci. Intell. Syst. **3**(1), 151–168 (2014)
7. Arunkumar, M., Karthikeyan, S.: Fuzzy Banach algebra stability of reciprocal quadratic functional equation via fixed point approach. Int. J. Pure Appl. Math. **119**(3), 31–39 (2018)
8. Bae, J.H.: On the stability of 3-dimensional quadratic functional equation. Bull. Korean Math. Soc. **37**(3), 477–486 (2000)
9. Bae, J.H., Park, W.G.: A functional equation originating from quadratic forms. J. Math. Anal. Appl. **326**, 1142–1148 (2007)
10. Bag, T., Samanta, S.K.: Fixed point theorems in Felbin type fuzzy normed linear spaces. J. Fuzzy Math. **16**(1), 243–260 (2008)
11. Bodaghi, A., Ebrahimdoost, Y.: On the stability of quadratic reciprocal functional equation in non-Archimedean fields. Asian-Eur. J. Math. **9**(1), 9 pages (2016)
12. Bodaghi, A., Kim, S.O.: Approximation on the quadratic reciprocal functional equation. J. Funct. Spaces Appl. Article ID532463, 5 pages (2014)
13. Bodaghi, A., Narasimman, P., Rassias, J.M., Ravi, K.: Ulam stability of the reciprocal functional equation in non-Archimedean fields. Acta Math. Univ. Comen. **85**(1), 113–124 (2016)
14. Bodaghi, A., Rassias, J.M., Park, C.: Fundamental stabilities of an alternative quadratic reciprocal functional equation in non-Archimedean fields. Proc. Jangjeon Math. Soc. **18**(3), 313–320 (2015)
15. Bodaghi, A., Senthil Kumar, B.V.: Estimation of inexact reciprocal-quintic and reciprocal-sextic functional equations. Mathematica J. **59**(1–2), No. 82, 3–14 (2017)

B. V. Senthil Kumar and H. Dutta, *Multiplicative Inverse Functional Equations*,
Studies in Systems, Decision and Control 289, https://doi.org/10.1007/978-3-030-45355-8

16. Cadariu, L., Radu, V.: Fixed points and the stability of quadratic functional equations. An. Univ. Timisoara Ser. Mat. Inform. **41**, 25–48 (2003)
17. Cadariu, L., Radu, V.: On the stability of the Cauchy functional equation: a fixed point apporach. Grazer Math. Ber. **346**, 43–52 (2004)
18. Chang, I.S., Jung, Y.S.: Stability of functional equations deriving from cubic and quadratic functions. J. Math. Anal. Appl. **283**, 491–500 (2003)
19. Chang, I.S., Kim, H.M.: On the Hyers-Ulam stability of a quadratic functional equations. J. Inequal. Appl. Math. **33**, 1–12 (2002)
20. Choi, M.D., Effros, E.: Injectivity and operator spaces. J. Funct. Anal. **24**, 156–209 (1977)
21. Chung, J., Kim, D., Rassias, J.M.: Hyers-Ulam stability on a generalized quadratic functional equation in distributions and hyperfunctions. J. Math. Phys. **50**(113519), 1–14 (2009)
22. Czerwik, S.: The stability of the quadratic functional equation. In: Rassias, T.M., Tabor, J. (eds.) Stability of Mappings of Hyers-Ulam Type, pp. 81–91. Hadronic Press, FL (1994)
23. Czerwik, S.: Functional Equations and Inequalities in Several Variables. World Scientific Publishing Company, NJ, London, Singapore and Hong Kong (2002)
24. Czerwik, S.: Stability of Functional Equations of Ulam-Hyers-Rassias Type. Hadronic Press, Palm Harbor, FL (2003)
25. Diaz, J., Margolis, B.: A fixed point theorem of the alternative for contractions on a generalized complete metric space. Bull. Am. Math. Soc. **74**, 305–309 (1968)
26. Ebadian, A., Zolfaghari, S., Ostadbashi, S., Park, C.: Approximation on the reciprocal functional equation in several variables in matrix non-Archimedean random normed spaces. Adv. Diff. Equ. **314**, 2015 (2015). https://doi.org/10.1186/s13662-015-0656-7
27. Effros, E.: On multilinear completely bounded module maps. Contemp. Math. **62**, 479–501 (1987)
28. Effros, E., Ruan, Z.J.: On matricially normed spaces. Pac. J. Math. **132**, 243–264 (1988)
29. Effros, E., Ruan, Z.J.: On approximation properties for operator spaces. Int. J. Math. **1**, 163–187 (1990)
30. Effros, E., Ruan, Z.J.: On the abstract characterization of operator spaces. Proc. Am. Math. Soc. **119**, 579–584 (1993)
31. Eskandani, G.Z., Rassias, J.M., Gavruta, P.: Generalized Hyers-Ulam stability for a general cubic functional equation in quasi-β-normed spaces. Asian-Eur. J. Math. **4**, 413–425 (2011)
32. Fassi, Z.I., Brzdek, J., Chahbi, A., Kabbaj, S.: On hyperstability of the biadditive functional equation. Acta Mathematica Scientia **37**(6), 1727–1739 (2017)
33. Fast, H.: Sur la convergence statistique. Colloq. Math. **2**, 241–244 (1951)
34. Felbin, C.: Finite dimensional fuzzy normed linear spaces. Fuzzy Sets Syst. **48**, 239–248 (1992)
35. Fridy, J.A.: On statistical convergence. Analysis **5**, 301–313 (1985)
36. Gajda, Z.: On the stability of additive mappings. Int. J. Math. Math. Sci. **14**, 431–434 (1991)
37. Gantner, T., Steinlage, R., Warren, R.: Compactness in fuzzy topological spaces. J. Math. Anal. Appl. **62**, 547–562 (1978)
38. Gavruta, P.: A generalization of the Hyers-Ulam-Rassias stability of approximately additive mapppings. J. Math. Anal. Appl. **184**, 431–436 (1994)
39. Gordji, M.E., Kaboli, G.S., Rassias, J.M., Zolfaghari, S.: Solution and stability of a mixed type additive, quadratic and cubic functional equation. Adv. Differ. Equ. Art ID 826130, 1–17 (2009)
40. Gordji, M.E., Zolfaghari, S., Rassias, J.M., Savadkouhi, M.B.: Solution and stability of a mixed type cubic and quartic functional equation in quasi-Banach spaces. Abstr. Appl. Anal. Art. 417473, 1–14 (2009)
41. Gordji, M.E., Gharetapeh, S.K., Park, C., Zolfaghri, S.: Stability of an additive-cubic-quartic functional equation. Adv. Differ. Equ. Art. 395693, 1–20 (2009)
42. Gordji, M.E., Savadkouhi, M.B., Park, C.: Quadratic-Quartic functional equations in RN-spaces. J. Inequal. Appl. Art. ID 868423, 1–14 (2009)
43. Gordji, M.E., Abbaszadeh, S., Park, C.: On the stability of a generalized quadratic and quartic type functional equation in quasi-Banach spaces. J. Inequal. Appl. Art. ID 153084, 1–26 (2009)

44. Gordji, M.E., Gharetapeh, S.K., Park, C., Zolfaghari, S.: Stability of an additive-cubic-quartic functional equation. Adv. Differ. Equ. Art. ID 395693, 1–20 (2009)
45. Gordji, M.E., Khodaei, H.: Solution and stability of generalized mixed type cubic, quadratic and additive functional equation in quasi-Banach spaces. Nonlinear Anal. **71**, 5629–5643 (2009)
46. Gordji, M.E.: Stability of a functional equation deriving from quartic and additive functions. Bull. Korean Math. Soc. **47**(3), 491–502 (2010)
47. Gordji, M.E., Khodaei, H., Khodabakhsh, R.: General quartic-cubic-quadratic functional equation in non-Archimedean normed spaces, U.P.B. Sci. Bull. Ser. A **72**(3), 69–84 (2010)
48. Gordji, M.E., Cho, Y.J., Ghaemi, M.B., Majani, H.: Approximately quintic and sextic mappings from r-divisible groups into Serstnev probabilistic Banach spaces: fixed point method. Discret. Dyn. Nat. Soc. Art. ID 572062, 1–16 (2011)
49. Gordji, M.E., Kamyar, M., Rassias, T.M.: General cubic-quartic functional equation. Abstr. Appl. Anal. Art. ID 463164, 1–18 (2011)
50. Gordji, M.E., Savadkouhi, M.B.: Stability of a mixed type additive, quadratic and cubic functional equation in random normed spaces. Filomat **25**(3), 43–54 (2011)
51. Hoseinia, H., Kenary, H.Z.: Stability of cubic functional equations in non-Archimedean normed spaces. Math. Sci. **5**(4), 321–336 (2011)
52. El-Hady, E.-S., Forg-Rob, W., Mahmoudi, M.: On a two-variable functional equation arising from databases. WSEAS Trans. Math. **14**, 265–270 (2015)
53. Hejmej, B.: Stability of functional equations in dislocated quasi-metric spaces. Ann. Math. Sil. **32**, 215–225 (2018)
54. Hensel, K.: Uber eine news Begrundung der Theorie der algebraischen Zahlen, Jahresber. Deutsch. Math. Ver. **6**, 431–436 (1897)
55. Hooda, N., Tomar, S.: Non-Archimedean stability and non-stability of quadratic reciprocal functional equation in several variables. Bull. Pure Appl. Sci. **37E**(2), 267–272 (2018)
56. Hooda, N., Tomar, S.: Approximation of reciprocal-cubic functional equation in non-Archimedean normed space. Int. J. Sci. Res. Math. Stat. Sci. **5**(5), 169–172 (2018)
57. Hyers, D.H.: On the stability of the linear functional equation. Proc. Nat. Acad. Sci. USA **27**, 222–224 (1941)
58. Hyers, D.H., Isac, G., Rassias, T.M.: Stability of Functional Equations in Several Variables. Birkhauser, Basel (1998)
59. Isac, G., Rassias, T.M.: Stability of ψ- additive mappings: applications to nonlinear analysis. Int. J. Math. Math. Sci. **19**(2), 219–228 (1996)
60. Javadi, S., Rassias, J.M.: Stability of general cubic mapping in fuzzy normed spaces. An. St. Univ. Ovidius Constanta **20**(1), 129–150 (2012)
61. Jin, S.S., Lee, Y.H.: On the stability of the quadratic-additive type functional equation in random normed spaces via fixed point method. Korean J. Math. **20**(1), 19–31 (2012)
62. Jun, K.W., Kim, J.M.: The generalized Hyers-Ulam-Rassias stability of a cubic functional equation. J. Math. Anal. Appl. **274**, 867–878 (2002)
63. Jun, K.W., Lee, S.B.: On the generalized Hyers-Rassias stability of a cubic functional equation. J. Chungcheong Math. Soc. **19**(2), 189–196 (2006)
64. Jung, S.M.: A fixed point approach to the stability of a Volterra integral equation. Fixed Point Theory Appl. Article ID 57064, 1–9 (2007)
65. Jung, S.M.: Hyers-Ulam-Rassias Stability of Functional Equations in Nonlinear Analysis. Springer, New York (2011)
66. Kaleva, O., Seikkala, S.: On fuzzy metric spaces. Fuzzy Sets Syst. **12**, 215–229 (1984)
67. Kannappan, P.: Theory of functional equations. Matsci. Rep. **48** (1969)
68. Kannappan, P.: Application of Cauchy's equation in combinatorics and genetics. Mathw. Soft Comput. **8**, 61–64 (2001)
69. Karakus, S.: Statistical convergence on probabilistic normed spaces. Math. Commun. **12**, 11–23 (2007)
70. Kolk, E.: The statistical convergence in Banach spaces. Tartu Ul. Toime. **928**, 41–52 (1991)

71. Kim, S.O., Senthil Kumar, B.V., Bodaghi, A.: Stability and non-stability of the reciprocal-cubic and reciprocal-quartic functional equations in non-Archimedean fields. Adv. Differ. Equ. **77**, 12 pages (2017)
72. Kuczma, M.: An Introduction to the Theory of Functional Equations and Inequalities. Panstwowe Wydawnictwo Naukowe-Uniwersytet Slaski, Warszawa-Krakow-Katowice (1985)
73. Lee, H., Kim, S.W., Son, B.J., Lee, D.H., Kang, S.Y.: Additive-quartic functional equation in non-Archimedean orthogonality spaces. Korean J. Math. **20**(1), 33–46 (2012)
74. Lee, J.R., Shin, D.Y., Park, C.: C. Hyers-Ulam stability of functional equations in matrix normed spaes. J. Inequal. Appl. (22), 1–11 (2013)
75. Margolis, B., Diaz, J.: A fixed point theorem of the alternative for contractions on a generalized complete metric space. Bull. Am. Math. Soc. **74**, 305–309 (1968)
76. Mirmostafaee, A.K.: Non-Archimedean stability of quadratic equations. Fixed Point Theory **11**, No. 1, 67–75 (2010)
77. Mohamadi, M., Cho, Y.J., Park, C., Vetro, P., Saadati, R.: Random stability of an additive-quadratic-quartic functional equation. J. Inequal. Appl. Art. ID 754210, 1–18 (2010)
78. Mohiuddine, S.A., Danish Lohani, Q. M.: On generalized statistical convergence in intuitionistic fuzzy normed space. Chaos Solitons Fractals **42**, 1731–1737 (2009)
79. Mohiuddine, S.A., Şevli, H.: Stability of Pexiderized quadratic functional equation in intuitionistic fuzzy normed space. J. Comput. Appl. Math. **235**, 2137–2146 (2011)
80. Moradlou, F., Rezaee, S., Sadeqi, I.: Approximate quadratic functional equation in Felbin's type normed linear spaces. Hacet. J. Math. Stat. **45**(3), 501–516 (2013)
81. Moradlou, F., Vaezi, H., Eskandani, G.Z.: Hyers-Ulam-Rassias stability of a quadratic and additive functional equation in quasi-Banach spaces. Mediterr. J. Math. **6**, 233–248 (2009)
82. Moslehian, M.S., Sadeghi, G.: Stability of two types of cubic functional equations in non-Archimedean spaces. Real Anal. Exch. **33**(2), 375–384 (2008)
83. Murali, R., Antony Raj, A.: Thamizharasan, T.: Stability of reciprocal difference and adjoint functional equations in multi-Banach spaces: a fixed point method. J. Emerg. Technol. Innov. Res. **5**(8), 450–455 (2018)
84. Mursaleen, M., Mohiuddine, S.A.: On stability of a cubic functional equation in intuitionistic fuzzy normed spaces. Chaos Solitons Fractals **42**, 2997–3005 (2009)
85. Mursaleen, M., Mohiuddine, S.A.: Statistical convergence of double sequences in intuitionistic fuzzy normed spaces. Chaos Solitons Fractals **41**, 2414–2421 (2009)
86. Mursaleen, M.: λ-statistical convergence. Math. Slovaca **50**, 111–115 (2000)
87. Mursaleen, M., Mohiuddine, S.A.: On lacunary statistical convergence with respect to the intuitionistic fuzzy normed space. J. Comput. Appl. Math. **233**, 142–149 (2009)
88. Mursaleen, M., Karakaya, V., Mohiuddine, S.A.: Schauder basis, separability, and approximation property in intuitionistic fuzzy normed space. Abstr. Appl. Anal. Art. ID 131868, 1–14 (2010)
89. Mursaleen, M., Mohiuddine, S.A., Edely, O.H.H.: On the ideal convergence of double sequences in intuitionistic fuzzy normed spaces. Comput. Math. Appl. **59**, 603–611 (2010)
90. Najati, A.: Hyers-Ulam-Rassias stability of a cubic functional equation. Bull. Korean Math. Soc. **44**(4), 825–840 (2007)
91. Najati, A., Eskandani, G.Z.: Stability of a mixed additive and cubic functional equation in quasi-Banach spaces. J. Math. Anal. Appl. **342**, 1318–1331 (2008)
92. Narasimman, P.: Solution and stability of a generalized kadditive functional equation. J. Interdiscip. Math. **21**(1), 171–184 (2018)
93. Narasimman, P., Dutta, H., Jebril, I.H.: Stability of mixed type functional equation in normed spaces using fuzzy concept. Int. J. Gen. Syst. **48**(5), 507–522 (2019)
94. Narasimman, P., Dutta, H.: $kn - 2$ scalar matrix and its functional equations by mathematical modeling. Adv. Top. Math. Anal. Chapter **14**, 449–468 (2019)
95. Park, C.: Fixed points and the stability of an AQCQ-functional equation in non-Archimedean normed spaces. Abstr. Appl. Anal. Art. ID 849543, 1–15 (2010)
96. Park, C., Shin, D.Y.: Functional equations in paranormed spaces. Adv. Differ. Equ. **123**, 1–23 (2012)

97. Park, C.: Stability of an AQCQ-functional equation in paranormed space. Adv. Differ Equ. **148**, 1–19 (2012)
98. Park, C., Shin, D.Y., Saadati, R., Lee, J.R.: A fixed point approach to the fuzzy stability of an AQCQ-functional equation. Filomat **30**(7), 1833–1851 (2016)
99. Park, J.H.: Intuitionistic fuzzy metric spaces. Chaos Solitons Fractals **22**, 1039–1046 (2004)
100. Park, K.H., Jung, S.: Stability of a cubic functional equation on groups. Bull. Korean Math. Soc. **41**(2), 347–357 (2004)
101. Park, W.G., Bae, J.H.: Solution of a vector variable bi-additive functional equation. Commun. Korean Math. Soc. **23**(2), 191–199 (2008)
102. Park, W.G., Bae, J.H.: Stability of a bi-additive functional equation in Banach modules over a C^*-algebra. Discret. Dyn. Nat. Soc. Article ID 835893, 1–12 (2012)
103. Pedrycz, W.: Some applicational aspects of fuzzy relational equations in system analysis. Int. J. Gen. Syst. **9**(3), 125–132 (1983)
104. Pedrycz, W.: Fuzzy relational equations: bridging theory, methodology and practice. Int. J. Gen. Syst. **29**(4), 529–554 (2000)
105. Perfilieva, I., Gottwald, S.: Solvability and approximate solvability of fuzzy relation equations. Int. J. Gen. Syst. **32**(4), 361–372 (2003)
106. Pinelas, S., Arunkumar, M., Sathya, E.: Hyers type stability of a radical reciprocal quadratic functional equation originating from 3 dimensional Pythagorean means. Int. J. Math. Appl. **5**(4), 45–52 (2017)
107. Pisier, G.: Grothendiecks theorem for non-commutative C^*-algebras with an appendix on Grothendiecks constants. J. Funct. Anal. **29**, 397–415 (1978)
108. Prager, W., Schwaiger, J.: Stability of the multi-Jensen equation. Bull. Korean Math. Soc. **45**(1), 133–142 (2008)
109. Rajkumar, V.: The generalized Hyers-Ulam-Rassias stability of a quadratic functional equation. Int. J. Res. Sci. Innov. **5**(7), 73–76 (2018)
110. Rassias, J.M.: On approximation of approximately linear mappings by linear mappings. J. Funct. Anal. **46**, 126–130 (1982)
111. Rassias, J.M.: Solution of a quadratic stability Hyers-Ulam type problem. Ricerche di Matematica, L, fasc. 1°, 9–17 (2001)
112. Rassias, J.M.: Solution of the Ulam stability problem for cubic mappings. Glasnik Matematicki Ser. III **36**(56), 63–72 (2001)
113. Rassias, J.M., Thandapani, E., Ravi, K., Senthil Kumar, B.V.: Functional Equations and Inequalities: Solution and Stability Results. World Scientific Publishing Company, Singapore (2017)
114. Rassias, T.M.: On the stability of the linear mapping in Banach spaces. Proc. Am. Math. Soc. **72**, 297–300 (1978)
115. Rassias, T.M.: Functional Equations, Inequalities and Applications. Kluwer Acedamic Publishers, Dordrecht, Bostan, London (2003)
116. Ravi, K., Arunkumar, M., Rassias, J.M.: On the Ulam stability for the orthogonally general Euler-Lagrange type functional equation. Int. J. Math. Sci. **3**(8), 36–47 (2008)
117. Ravi, K., Rassias, J.M., Pinelas, S., Narasimman, P.: The stability of a generalized radical reciprocal quadratic functional equation in Felbins space. PanAmerican Math. J. **24**(1), 75–92 (2014)
118. Ravi, K., Rassias, J.M., Senthil Kumar, B.V.: A fixed point approach to the generalized Hyers-Ulam stability of reciprocal difference and adjoint functional equations. Thai J. Math. **8**(3), 469–481 (2010)
119. Ravi, K., Senthil Kumar, B.V.: Ulam-Găvruta-Rassias stability of Rassias reciprocal functional equation. Glob. J. Appl. Math. Math. Sci. **3**(1–2), 57–79 (2010)
120. Ravi, K., Rassias, J.M., Narasimman, P.: Stability of a cubic functional equation in fuzzy normed space. J. Appl. Anal. Comput. **1**(3), 411–425 (2011)
121. Ravi, K., Senthil Kumar, B.V.: Stability and geometrical interpretation of reciprocal functional equation. Asian J. Curr. Engg. Math. **1**(5), 300–304 (2012)

122. Ravi, K., Senthil Kumar, B.V.: Generalized Hyers-Ulam Stability of a system of bi-reciprocal functional equations. Eur. J. Pure Appl. Math. **8**(2), 283–293 (2015)
123. Ravi, K., Suresh, S.: Generalized Hyers-Ulam stability of a cubic reciprocal functional equation. Br. J. Math. Comput. Sci. **20**(6), 1–9 (2017)
124. Rodabaugh, S.E.: Fuzzy addition in the L-fuzzy real line. Fuzzy Sets Syst. **8**, 39–51 (1982)
125. Ruan, Z.J.: Subspaces of C^*-algebras. J. Funct. Anal. **76**, 217–230 (1988)
126. Saadati, R., Park, J.H.: On the intuitionistic fuzzy topological spaces. Chaos Solitons Fractals **27**, 331–344 (2006)
127. Saadati, R., Cho, Y.J., Vahidi, J.: The stability of the quartic functional equation in various spaces. Comput. Math. Appl. **60**, 1994–2002 (2010)
128. Saadati, R., Park, C.: Non-archimedean L-fuzzy normed spaces and stability of functional equations. Comput. Math. Appl. **60**, 2488–2496 (2010)
129. Sadeqi, I., Moradlou, F., Salehi, M.: On approximate Cauchy equation in Felbin's type fuzzy normed linear spaces. Iran. J. Fuzzy Syst. **10**(3), 51–63 (2013)
130. Sahoo, P.K., Kannappan, P.l.: Introduction to Functional Equations. CRC Press, Taylor & Francis Group (2011)
131. Salat, T.: On the statistically convergent sequences of real numbers. Math. Slovaca **30**, 139–150 (1980)
132. Senthil Kumar, B.V., Bodaghi, A.: Approximation of Jensen type reciprocal functional equation using fixed point technique. Boletim da Sociedade Paranaense de Matematica (Article in press)
133. Senthil Kumar, B.V., Dutta, H.: Non-Archimedean stability of a generalized reciprocal-quadratic functional equation in several variables by direct and fixed point methods. Filomat **32**(9), 3199–3209 (2018)
134. Senthil Kumar, B.V., Dutta, H.: Fuzzy stability of a rational functional equation and its relevance to system design. Int. J. Gen. Syst. **48**(2), 157–169 (2019)
135. Senthil Kumar, B.V., Dutta, H.: Approximation of multiplicative inverse undecic and duodecic functional equations. Math. Meth. Appl. Sci. **42**, 1073–1081 (2019)
136. Shin, D., Lee, S., Byun, C., Kim, S.: On matrix normed spaces. Bull. Korean Math. Soc. **27**, 103–112 (1983)
137. Shukla, S., Gopal, D., Roldán-López-de-Hierro, A.-F.: Some fixed point theorems in 1-M-complete fuzzy metric-like spaces. Int. J. Gen. Syst. **45**, 7–8, 815–829 (2016)
138. Skof, F.: Proprietá locali e approssimazione di operatori. Rend. Semin. Mat. Fis. Milano **53**, 113–129 (1983)
139. Song, A., Song, M.: The stability of quadratic-reciprocal functional equation. In: AIP Conference Proceedings, vol. 1955, p. 040171 (2018). https://doi.org/10.1063/1.5033835
140. Steinhaus, H.: Sur la convergence ordinaire et la convergence asymptotique. Colloq. Math. **2**, 73–34 (1951)
141. Tao, N., Zhu, Y.: Stability and attractivity in optimistic value for dynamical systems with uncertainty. Int. J. Gen. Syst. **45**(4), 418–433 (2016)
142. Ulam, S.M.: Problems in Modern Mathematics, Chapter VI. Wiley-Interscience, New York (1964)
143. Wilansky, A.: Modern Methods in Topological vector space. McGraw-Hill International Book Co., New York (1978)
144. Wiwatwanich, A., Nakmahachalasint, P.: On the stability of a cubic functional equation. Thai J. Math. Spec. Issue (Annual Meeting in Mathematics), 69–76 (2008)
145. Xia, M.: Interval-valued intuitionistic fuzzy matrix games based on Archimedean t-conorm and t-norm. Int. J. Gen. Syst. **47**(3), 278–293 (2018)
146. Xiao, J., Zhu, X.: On linearly topological structure and property of fuzzy normed linear space. Fuzzy Sets Syst. **125**, 153–161 (2002)
147. Xiao, J., Zhu, X.: Topological degree theory and fixed point theorems in fuzzy normed space. Fuzzy Sets Syst. **147**, 437–452 (2004)
148. Xu, T.Z., Rassias, J.M., Xu, W.X.: Intuitionistic fuzzy stability of a general mixed additive-cubic equation. J. Math. Phys. **063519**(51), 1–21 (2010)

149. Xu, T.Z., Rassias, J.M., Xu, W.X.: Generalized Ulam-Hyers stability of a general mixed AQCQ-functional equation in multi-Banach spaces: a fixed point approach. Eur. J. Pure Appl. Math. **3**(6), 1032–1047 (2010)
150. Zadeh, L.A.: Fuzzy sets. Inform. Control **8**, 338–353 (1965)

Printed in the United States
by Baker & Taylor Publisher Services